Praise for *Living the 1.5* [illegible]

For years I've relied on Lloyd Alter's [illegible] insight in *Tree-hugger*, including his direct, and to-th[illegible] point reflections on the elements of a better urban life. In his new, equally clear book, he's done it again with an immensely practical set of insights and rules to live by. With wisdom gleaned during the pandemic-year, Lloyd joins Paul Hawken as a master of the carbon drawdown roadmap.

— Chuck Wolfe, Seeing Better Cities Group,
author, *Sustaining a City's Culture and Character*

This joyful adventure in 1.5 degree living shows that a focus on "sufficiency" can unlock a happier, healthier sustainable way of life. "Efficiency" has dominated precisely because it perpetuates the *status quo* but Lloyd´s story shines a light on why we must transform our vision of a "good life" as well as practical tips for how to do it.

— Kate Power, development director, Hot or Cool Institute

In the race against climate change some lifestyle changes make a big difference and some are negligible. Lloyd Alter's on-point analysis helps us sort out the differences. Consumers must reduce demand for climate-cooking goods and services, but policy changes are equally needed to remove institutional barriers to sustainable ways of life. Alter shows us that we need to re-learn what it is that we really *need* for prosperity, and unlearn the unquenchable thirst for "more," which the infinite growth economy has worked so hard to instill.

— Bart Hawkins Kreps, writer, editor

Lloyd Alter's writing is always fresh, incisive, and thoughtful and, in this timely and clearly written call to action, he tells us exactly what we as citizens of the world must do in our everyday lives to reduce our greenhouse gas emissions before it's too late. Reflecting Alter's background as an architect and real estate developer, this book is at once idealistic and pragmatic, and the lifestyle issues it addresses—based on the author's own experience—could not be more urgent. Governments and businesses cannot solve this problem without actions by each of us as individuals, and Alter tells us how.

— F. Kaid Benfield, Senior Counsel, PlaceMakers LLC

The toughest question for the next two decades is how to pull back from the brink. Part of the answer has to do with government policy and low-carbon technologies. But the path to net zero has to be about more than just consenting to new rules. It must also be about figuring out how we can reduce carbon through our choices: where we live, what we eat, how we move around, and so on.

— John Lorinc, writer, journalist

With fact-based research and tales of personal experience, Lloyd will reach many new audiences with this book. I, for one, embrace a return to the concept of sufficiency with regard to transportation, food, and more. For all who've been preaching climate action, or skirting it, I implore you to read this book and own up to all the ways you could try harder, and then get louder about sharing the overall life benefits you gain.

— Andrea Learned, climate leadership strategist,
founder, #Bikes4Climate

Lloyd Alter has created a highly engaging, intelligent, and thoughtful "manual for living the 1.5-degree lifestyle." There is plenty of myth-busting, supported by quality, in-depth research which goes far beyond the usual well-worn clichés and mantras. He humanises and crystallizes the "choices and trade-offs" we have to make to keep within 1.5°C global warming, so "it is sufficient with enough to go around for everyone."

— Rosalind Readhead, English climate activist

Living the 1.5 Degree Lifestyle

Why Individual Climate Action Matters More than Ever

- Lloyd Alter -

Cover design by Diane McIntosh.
Cover image: ©iStock.

Printed in Canada. First printing September, 2021.

Any other inquiries can be directed by mail to:

New Society Publishers
P.O. Box 189, Gabriola Island, BC V0R 1X0, Canada
(250) 247-9737

LIBRARY AND ARCHIVES CANADA CATALOGUING IN PUBLICATION

Title: Living the 1.5 degree lifestyle : why individual climate action matters more than ever / Lloyd Alter.

Other titles: Living the one point five degree lifestyle
Names: Alter, Lloyd, author.

Description: Includes bibliographical references and index.

Identifiers: Canadiana (print) 20210194626 | Canadiana (ebook) 20210194820 | ISBN 9780865719644 (softcover) | ISBN 9781550927573 (PDF) | ISBN 9781771423533 (EPUB)

Subjects: LCSH: Sustainable living. | LCSH: Environmental protection—Citizen participation. | LCSH: Climate change mitigation—Citizen participation.

Classification: LCC GE196 .A48 2021 | DDC 640.28/6—dc23

Funded by the Government of Canada | Financé par le gouvernement du Canada | Canada

New Society Publishers' mission is to publish books that contribute in fundamental ways to building an ecologically sustainable and just society, and to do so with the least possible impact on the environment, in a manner that models this vision.

Contents

Dedication

To my wife, Kelly, who put up with all of this,
and to kids, Emma, who rides a bike all winter,
and Claire, who never learned to drive.
I evidently had some small influence.

Acknowledgments

First, I have to acknowledge Rosalind Readhead in London, who did this before me and turned me on to the idea. But then I have to go back to Graham Hill, who started *Treehugger* and gave me a ten-buck-a-post gig writing 15 years ago. That eventually led to a working full-time with a succession of really smart people, including Meaghan O'Neill, Ken Rother, Emily Murphy, Molly Fergus, and Melissa Breyer who put up with me and edited me. Then there were people who inspired and influenced me, including Steve Mouzon, Kaid Benfield, Alex Steffen, and Kris de Decker. But perhaps the most important event was learning about Passivhaus and all the people in the Passivhaus community that I have become friends with and learned so much from, including Wolfgang Feist, Monte Paulsen, Mike Eliason, Bronwyn Barry, Elrond Burrell, Ken Levenson, and so many I have missed here. Also pivotal was teaching at the Ryerson School of Interior Design; Chair Lois Weinthal and a dozen classes of students who taught me more than I ever taught them. Finally, there are the people who got me to this stage with this book, including lawyer Willa Marcus and editors Rob West and Judith Brand.

The 1.5-Degree Lifestyle: Introduction

I used to have a monster carbon footprint. At the end of the last century, I was in my second career (my first was as an architect) as a successful real estate developer in Toronto, building award-winning condominiums. I drove my classic Porsche 914 the couple of blocks between office and jobsite; I drove my daughters to school and down to the lake every morning in the rowing season; then on winter weekends, we drove up to the private ski club where all the rich developers hung out. Every weekend in summer, I drove the family up to our summer cottage in Muskoka. Throw in a few flights every year, and I was probably emitting about 30 tonnes of CO_2 per year in the process, or what could be called a 30-tonne lifestyle.

Then, suddenly, I wasn't a developer anymore; after a falling-out with partners, I had almost nothing but a substantial financial loss and probably a nervous breakdown. However, I had learned a great deal from the experience; I was convinced that the way we build had to change, that it was too slow and too expensive and used too much material and energy. I went to the biggest prefabricated housing manufacturer in the province and convinced him to let me design and sell small, modern, green housing units. He agreed, I set up an office, started doing all the home shows, and waited for the phone to ring, when I wasn't driving my Subaru all over the province. While waiting, I built a website to educate people about prefab and green building, updating it every day as I would find articles of interest—essentially a blog before there were blogs.

I spent a lot of time waiting for that phone to ring; there wasn't much interest in Ontario in small modern green prefab. However, in the United States, there was huge interest in my website, which was soon recognized as one of the most important resources on prefab in the time before blog platforms appeared in about 2004. One of the first that I started following was a website called *Treehugger*, which was then a "guide to green and gorgeous." I started sending them tips, stuff that I couldn't use on my own business-related site. Soon I was writing for them as well for $10 a post, and not long after that was offered a full-time position. I concluded that I was a better writer than I was a prefab salesman and have been doing it ever since.

There were other changes; the Chair of the Ryerson School of Interior Design saw me speak on a panel and asked me to apply for an open position teaching sustainable design. The more I read, the more I taught, and the more I wrote, the more concerned I became about the issues of sustainability. My carbon footprint was dropping because I couldn't afford that developer lifestyle anymore, but also because I was becoming increasingly concerned about the issues. Having built out of both concrete and wood, I became an early proponent of the concept of embodied carbon, which almost nobody took seriously 10 years ago. (Actually, they still don't.) My carbon footprint might have been even lower had not my early focus on wood, embodied carbon, and an efficient building concept called Passive House put me on the international lecture circuit stretching from Seattle to Munich, or my press-related trips to China and Spain.

Much changed with the Paris Agreement in 2015, with its limits on carbon emissions. According to the Intergovernmental Panel on Climate Change (IPCC), we have to cut the quantity of greenhouse gases we emit roughly in half by 2030, and almost to zero by 2050, if we want to keep the rise in global average temperature to 1.5 degrees Celsius and avoid catastrophic consequences of global warming. To put this in perspective, the COVID-19 lockdown, with its massive reductions in transportation and industry, reduced emissions by about 7%. We have to continue doing that every year, another 7% to 8% reduction, to stay under 1.5 degrees.

But where do these greenhouse gas emissions come from? Who is responsible? Who has to fix it? How can we fix it? Suddenly the measuring of all the carbon that we are all putting into the air is of critical importance, and someone has to fix it or be blamed for it.

Everyone has heard the statement that "100 companies are responsible for 71% of global emissions," that corporations emit the carbon and governments should regulate the carbon away. They should fix the problem by delivering clean electricity instead of burning fossil fuels, or by running our pickup trucks on electricity. The latest is to put hydrogen in our furnaces instead of gas, so that we can all keep living the way we do until we have to maybe start thinking about this in 2030.

The problem with this is that those 100 companies don't directly produce much CO_2; they sell fossil fuels that are burned for energy, which releases CO_2. It's their customers, you and me, who turn their product into emissions. We buy what they are selling, directly or indirectly, whether out of choice or out of necessity.

Most of the world's nations signed on to the Paris Agreement, promising to reduce their carbon emissions, but so far nobody has done very much. It's hard when you have economies based on digging up fossil fuels and then manufacturing stuff that runs on them, emitting carbon at every step of the way. It's harder when everyone wants more stuff, and the jobs all depend on us buying it. So, the only strategy anyone can think of is to produce more carbon-efficient stuff, to build electric cars instead of gasoline-powered, to burn natural gas instead of coal, to make more wind turbines and solar panels, and to dream of nuclear reactors, carbon capture and storage, and hydrogen.

This was actually working, to a degree: pre-pandemic, the rate of increase in carbon emissions was slightly less than the growth of the world's economies. But even with all that greening going on, carbon emissions were still increasing by 1.3% on average, while the global economy expanded by about 3%.[1] And in 2019, global greenhouse gas emissions from all sources still reached a record high of 52.4 gigatonnes of CO_2e. (The e stands for equivalents—other gases like methane or fluorocarbon refrigerants, some of which have many thousands of times the global warming potential of CO_2.) When the economy booms, so do emissions.

The world loves growth, and nobody wants to see an economic seizure like we had during the pandemic happen again. Governments have been pouring vast sums into cranking up the economic engines, encouraging us to buy more stuff and more services, while almost completely ignoring the fact that to keep under a temperature rise of 1.5 degrees, we have to reduce our carbon emissions budget to 25 gigatonnes of CO_2e by 2030, less than half of what we emitted in 2019.

Norman Mailer wrote, "There was that law of life, so cruel and so just, that one must grow or else pay more for remaining the same." Growth is the law of life, and the engine of growth runs on fossil fuels.

If we have any chance of getting close to the carbon budget for 2030, we have to change the way we think about growth. We have to stop thinking about production, the making of what everyone is selling, and start thinking about consumption, what we are buying. We have to stop thinking about efficiency, making something slightly better, and start thinking about sufficiency: what do we really need?

The premise of this book, and the research it is based on, is that we are all collectively responsible for reducing our carbon emissions to keep under that 1.5-degree ceiling. We have that carbon budget set in Paris, and if you divide it by the number of people on Earth, we have a personal carbon allocation or budget target of "lifestyle emissions," those emissions that we can control, of about 2.5 tonnes per person, per year by 2030. Getting by on this is what we are calling the 1.5-degree lifestyle.

But what is living on 2.5 tonnes of carbon actually like? How do you measure it? How much does individual consumption matter? These are some of the questions that this book will try to answer.

We will try and look at the carbon cost of everything that we do in our lives to help people make choices about what makes sense, what's worth trying to change, and what isn't. It's a model that not only can influence our personal lives but also can guide policy, from urban planning to agriculture.

For many people, lifestyle carbon emissions are baked into the way we live and very hard to change without concomitant societal and environmental changes; our developed Western world seems almost designed to emit carbon. We are also creatures of habits that are difficult to shake. However, many habits changed in the course of the COVID-19 pandemic. It was perhaps not the best time to start this journey; much of the planet was now living a low-carbon lifestyle whether they wanted to or not.

On the other hand, it may be the perfect time for changes. We can collectively work for system change, but also for individual change, a 1.5-degree lifestyle. It is based on living within a tight carbon budget, but if one makes the right choices, it is sufficient, and there is enough to go around for everyone.

— 1 —

What's the 1.5-Degree Lifestyle?

- 1 grapefruit: 90 grams of carbon
- Instant coffee with milk: 50 g
- 1-mile cycle: 3 g
- Seared mackerel fillet with British seasonal asparagus and Jersey new potatoes: 600 g
- Toasted hazelnuts, honey, and yogurt: 200 g
- 1 orange: 90 g
- Eggy bread (1 egg = 300 g) and kimchi: 400 g
- Cardamom and honey milk: 350 g
- Time online approximately 3 hours
- Data and servers: 3 × 50: 150 g
- Device laptop iPad or iPhone: used 3 times
- Fridge: 64 g
- Average water use: 38 g

That's a day in the life of Rosalind Readhead[1] (May 14, 2020 to be more precise), an English activist and erstwhile mayoral candidate, expressed in grams of carbon emissions. She measures every move that she makes, every bite that she eats, everything that she does, in her attempt to live a lifestyle with emissions totaling less than 1 tonne per year or 2.74 kilograms of CO_2 or its equivalents per day.

Rosalind is living her radical version of the 1.5-degree lifestyle, a demonstration of how we have to live to meet the target set by the Paris agreement on climate change if we are going to keep the global temperature rise below 1.5 degrees Celsius. As noted above, to get there, we have to reduce the average carbon footprint of each person on the planet to 2.5 tonnes by 2030, and then further reduce this to 1 tonne by 2050.

Rosalind is trying to live with the 2050 target of 1 tonne. That is almost impossible in today's society; for most people, it is almost a baseline of the stuff that they can't change or avoid. In the longer term, it is achievable after we rebuild our homes, rethink our offices, and reimagine our lifestyles. I have been trying to live a reduced carbon lifestyle as well, but have been aiming for the far less onerous 2030 target of 2.5 tonnes of carbon per year.

Either consciously or by accident, I have already made lifestyle choices that make it easier. I live in a province of Canada that has low-carbon electricity from nuclear and hydropower, in a streetcar suburb where I can get almost everything I need without driving. I work from home. For others, it's not so simple.

Nonetheless, this book is an attempt at a manual for living the 1.5-degree lifestyle, looking at the choices and trade-offs that we have to make to get there. Perhaps more usefully, it is a look at where our carbon emissions come from and how we got into this mess in the first place.

Some will find it harder than others; some might find it impossible without drastic changes and serious investments.

Many will say, why bother? Everybody knows that it is governments, oil companies, and industry that cause the CO_2 emissions, that our individual actions don't matter or won't make a difference. Others say that we have to get out in the streets and fight for system change, for regulatory change, and for government change. In fact, we need all of the above. But one only needs to look back at the lockdown in spring 2020; we saw what happens when a lot of people stop driving and flying. All the associated industries almost collapsed because the demand for their products and services disappeared overnight. None of it was by choice, but it proved that these businesses are like any other, demand-driven. If we don't buy what they are selling, then they have to change or go under.

What big business wants you to do is buy their goods and services, which generates the emissions. Some of this is by choice, and when we choose not to buy, we are not only emitting less carbon, we are also emitting less money—a low-carbon lifestyle is generally cheaper. It's also healthier, leading to a better diet and more exercise.

The 1.5-degree lifestyle is not only good for the planet, it's good for you.

What's So Special about 1.5 Degrees?

Buckminster Fuller once asked a very young Norman Foster, "How much does your building weigh?" Usually in architecture, the only person who cares about this is the engineer designing the foundations, part of a building that is buried in the ground and never seen again. But it is critical; it determines whether a building stays up or falls over. Numbers matter.

I am an architect and a writer, not a climate scientist, so I do not want to get into the details of what is causing climate change. I am assuming that people who are reading this already know, but if not, lots of hefty scientific reports from the IPCC and others and many terrific recent books do this very well, like Eric Holthaus's *The Future Earth* or Peter Kalmus's *Being the Change*, which expresses his very personal and somewhat emotional point of view. I am not an emotional person (my wife and kids will confirm this); I like numbers. Targets. Things I can measure, weigh, quantify. I really like spreadsheets, all those numbers laid out for anyone to see. When my late father retired many years ago, I got him a PC with Lotus 123 on it; he would deconstruct corporate financial statements for fun and could proudly tell you how much change he had in his pocket by looking at his screen. Perhaps I got it from him.

That's probably why I am so attracted to the 1.5-degree lifestyle; I can measure this. Sort of. But I will try to explain where the numbers come from.

The 1.5-degree *target* comes from the 2015 Paris Agreement, where nations committed to "holding the increase in the global average temperature to well below 2°C above preindustrial levels and pursuing efforts to limit the temperature increase to 1.5°C above preindustrial levels." The Paris Agreement is based on the scientific consensus, but it is a policy document, a treaty, that includes numbers that the signatories can verify: a target (the temperature), a path to get to the target (the carbon budget), and a schedule.

The first challenge was to figure out what happens at various degrees of warming, and what target to aim for. Using historical data and "multiple forms of knowledge, including scientific evidence, narrative scenarios and prospective pathways," the IPCC estimated the effects on the climate from warming at various temperatures. The 2015 Agreement settled on 2°C as the target, but in 2018 the IPCC released a special report that showed what a difference half a degree makes: 2.6 times as many extreme heat events, twice as much species loss, reductions in crop yields by half.

Half a degree doesn't sound like much, but we are not starting our measurements now, but back at the preindustrial levels of 1880, and we are already at about 1 degree, so from our standing start now, it is double the temperature rise. It is a big enough difference that there was concern about "tipping points"[2] that occurred below 2 degrees, with particular worry about sea ice and permafrost collapse. But even at 1.5 degrees, we've got trouble.

Even at 1.5 degrees, we face extreme changes, with more extremely hot days, droughts in some areas, and heavy rains in others. There will be impacts on biodiversity and ecosystems: in the north, the transformation of the boreal forests, tundra, and permafrost; changes to the ranges of marine species and reduced productivity of fisheries.

Even at 1.5 degrees, human health is affected by heat-related mortality, by increased air pollution, by geographic spread of diseases like malaria and dengue fever.

Even at 1.5 degrees, there may be reductions in crop yields and food availability. Livestock may also be challenged by changes in food supply and disease.

The Carbon Budget

The *budget* starts with the "simple idea," as Zeke Hausfather of Carbon Brief calls it, that "the amount of global surface temperature warming tends to increase proportionately with the total cumulative emissions of CO_2."[3] The next challenge was to calculate how much could be emitted before the specified temperature rise was likely to occur. This was, again, complicated science given that the oceans and forests absorb so much carbon and so many variables have to be separated out to determine the quantity of anthropogenic emissions. As Bard Lahn notes, it is "a concept explicitly aimed at mediating between scientific knowledge and policymaking."[4] He continues in his "History of the Global Carbon Budget":

> Alongside its scientific merits, therefore, the main strength of the carbon budget concept was seen by scientists to lie in its ability to simplify and accentuate certain choices and challenges facing policymakers. It was based on this line of reasoning that the IPCC AR5 report unequivocally concluded that "the simplicity of the concept of a cumulative carbon emission budget makes it attractive for policy."[5]

The carbon budget measures CO_2e, or carbon dioxide equivalents. CO_2 is not the only greenhouse gas; others include methane and the fluorocarbons like refrigerants. Some have many times the effect of CO_2 and are converted into equivalents based on their Global Warming Potential (GWP); methane, for example, has a GWP 25 times that of CO_2, so 1 kilo of methane is counted as 25 kilos of CO_2. Time is also a factor; CO_2 stays in the atmosphere almost forever, but methane breaks down over about 20 years.

Finally, there is the *schedule*, with the long-range target of the end of the century, a mid-range target of 2050, and staring us in the face, the short-term target of 2030. To keep the temperature rise under 1.5 degrees at the end of the century, we cannot have cumulative emissions of more than 420 gigatonnes; according to the latest UN Environmental Programme emissions gap report, global greenhouse gas emissions in 2019 were 52.4 gigatonnes. To stay under 1.5 degrees, we have to start reducing that to 25 gigatonnes by 2030 (less than half of what they were before the pandemic) and essentially to net zero by 2050.

Meanwhile, back in the real world, emissions were rising at 1.4% per year prior to 2020. That is the scale of the challenge we face: we have to seriously, dramatically, radically, and painfully reduce the amount of greenhouse gases we emit to keep under that carbon budget. That's a global drop of 7.9% per year. That doesn't sound so dire, until you realize that the COVID-19 pandemic—when factories closed, flights were grounded, and nobody was driving, an almost total shutdown of the global economy—is estimated to have caused a global drop of CO_2 emissions of about 7%.

It also doesn't mean that we can all talk about this until 2030, we have to start now. Climate scientist Kate Marvel said it best (three years ago!):

> You may have heard that we have 12 years to fix everything. This is well-meaning nonsense, but it's still nonsense. We have both no time and more time. Climate change isn't a cliff we fall off, but a slope we slide down. And, true, we've chosen to throw ourselves headlong down the hill at breakneck speed. But we can always choose to begin the long, slow, brutal climb back up.[6]

-2-

Equity, Fairness, and the 2.5-Tonne Budget

The world has a carbon budget for 2030 and 2050, as do nations in their Nationally Determined Contributions submissions that are part of the Paris Agreement. Individuals do not, and they vary widely; the average per capita consumption emissions for an American are about 17.6 tonnes per year, while an average Indian emits only 1.7 tonnes. Meanwhile, the richest one percent of the world may have an annual footprint as high as 75 tonnes of "lifestyle" emissions, or those emissions directly attributable to what we as individuals do and how we live.

Max Roser of *Our World in Data* points out that half the world is emitting far too much carbon, but that the other half suffers from energy poverty: "Those that do not have sufficient access to modern energy sources suffer poor living conditions as a result."[1] Any fair and equitable division of the carbon budget has to allow headroom for those suffering from energy poverty to get a little more of it.

At the other end of the spectrum, when I fly to Portugal or drive my Subaru, I may get the benefit and pay the cost in dollars, but everyone in the world is affected by the carbon emissions. So a logical, equitable, and reasonable place to start is with an average carbon budget for everyone on the planet.

Lifestyle emissions are not just individual but the things that we share a piece of, from how we organize our society and our institutions. They are a big chunk of global emissions; a 2009 study concluded that "on the global level, 72% of greenhouse gas emissions are related to household consumption, 10% to government consumption, and 18% to investments."[2]

The next bit of math is also straightforward; we have a carbon emissions budget target of 25 gigatonnes in 2030 to stay under 1.5 degrees of warming. If you divide that by the world's population, the result is roughly 3.4 tonnes per person per year. Multiply that by 72% and you get a 2030 target lifestyle footprint of 2.5 tonnes of CO_2e per person per year. That's the 1.5-degree lifestyle.

Many will argue that expecting someone in the United States to lower their consumption to a worldwide average is crazy socialist talk, and that it will never happen. They are probably right, but it is a place to start. After all, we are not talking about money or status here, we are talking about carbon. The rich man can park his Tesla under his Tesla Solar Roof and charge his Tesla battery and have a very expensive but low operating carbon lifestyle. And frankly, it is not an unrealistic or unreasonable target.

I learned of the actual term "1.5-degree lifestyle" from Rosalind Readhead, who pointed me to a study from the Institute for Global Environmental Studies (IGES), Aalto University, and D-mat titled *1.5-Degree Lifestyles: Targets and Options for Reducing Lifestyle Footprints*. It provided the fundamental underpinning of this project; as noted in the introduction:

> Lifestyles of individuals consist of various elements of daily living including consumption relating to nutrition, housing, mobility, consumer goods, leisure, and services. The consumption-based accounting adopted in this study attributes GHG emissions at production stages as indirect emissions caused by household consumption. This provides a different angle from the footprint of specific products, organizations, cities, or countries, which have been the foci of most footprint studies so far.[3]

The lifestyle study authors acknowledge that this cannot be achieved by individuals on their own; much of it is structural and locked-in. Our world is designed around consumption of energy, and it is hard to break this pattern.

> Although this study quantifies impacts of GHG emissions from perspective of lifestyles and consumption by households, it does not mean that individual households are solely responsible

> for reducing the footprints. The sheer magnitude of change required for a shift towards 1.5-degree lifestyles can only be achieved through a combination of system-wide changes and a groundswell of actions from individuals and households.[4]

So much of our consumption is "baked in" to the way our economies are set up; we still need political action and societal change. But that doesn't give us carte blanche to blame the system and not take personal responsibility.

The Lifestyle Domains

The 1.5-degree lifestyle report studied people's lives in great detail in four countries, looking at six "lifestyle domains": nutrition, housing, mobility, consumer goods, leisure, and services. After studying the results in all six sectors, the authors concluded that about 75% of the impact fell within the hotspots of nutrition, housing, and mobility, basically what we eat, where we live, and how we get around.

I was not convinced of this; in my own situation, I have found that "communication and information" in the services category are in fact one of my biggest sources of emissions because I spend all day on my computer connected to the internet. My consumption of expensive Apple consumer goods turns out to eat up a lot of my carbon budget too, so we will look at all six sectors not just the hotspots.

The divisions are also somewhat arbitrary. It is also not so simple to think of them as six separate categories. I will show that housing and transportation are two sides of the same coin and that nutrition is affected by both, as are consumer goods. The North American family tends to drive an SUV to the big-box store once a week for groceries, putting much of them in a giant fridge. The urban Italian might have a tiny fridge, picking up the fresh fixings for dinner on the way home. The Japanese worker might get off the subway and find themselves surrounded by vast multi-level supermarkets. They don't buy the giant tub of ice cream; it's heavy, it might melt, and they don't have a freezer big enough for it. So dividing everything into six lifestyle domains is not really accurate; it's all connected and interrelated, and it is all a rough approximation.

However, the six lifestyle domains are a good place to start, a way to break things down into measurable categories.

Three Approaches to Reducing Our Footprints

The study authors describe three approaches we can take in each category to reduce our footprints, an extremely useful division.

Absolute Reduction

Absolute reduction "means reducing physical amounts of goods or services consumed, such as food, kilometers driven, energy use, or living space, as well as avoiding unsustainable options."

Simply put, just using less. This is the "less is more" and "living with less" approach that I have called *sufficiency*, asking the question How much do we really need?

Efficiency Improvement

Efficiency improvement "means decreasing emissions by replacing technologies with lower-carbon ones while not changing the amount consumed or used, such as in energy-efficient agriculture, vehicles, or housing."

This has always been the standard approach, improving the efficiency of everything that we make and use. But it has failed us: as cars got more efficient, they turned into SUVs; as houses got more efficient, they got bigger.

Modal Shift

Modal shift "means changing from one consumption mode to a less carbon intensive one, such as in adopting plant-based diets, using public transport, or renewable energy for electricity or heating."

This is perhaps the most interesting and important approach: doing things differently. Like absolute reduction, it is closely related to the concept of sufficiency: why drive a car when you can ride a bike, or why use a dryer when you can string a clothesline? Modal shifts also give us the greatest carbon emission reductions and the greatest opportunities.

Doing the Math

The challenge here is to live a lifestyle that emits less than 2.5 tonnes of carbon dioxide or CO_2 equivalents per year; the worldwide average is 4.8 tonnes, although there is not much point in comparing per capita emissions from production. Why are Chinese emissions per capita nearly four times as much as India's when they are both populous, rapidly developing countries? It's because the numbers are based on dividing the emissions produced by each nation by the population, and China is making so much stuff that we consume. Really, they are *our* emissions that we have offshored to China. Why are Canadian emissions lower than US emissions? The energy mix is cleaner, with more hydroelectric power. No thanks at all to the greener habits of Canadians.

On the other side of the ledger, we have to estimate the carbon footprint of what we do and what we eat. I learned from Rosalind Readhead about Mike Berners-Lee's 2011 book *How Bad Are Bananas*, which tried to put a real number on many items found in our everyday lives. But he admits right up front that it is a rough guide: "The carbon footprint, as I have defined it, is the climate change metric that we need to be looking at. The problem is that it is also impossible to measure." He also admits that some of his numbers are flaky: "Sometimes my calculations and assumptions are highly debatable, but I've included them because I think that just going through the thought process can be a useful reflection on something that matters."[5] But it is still a worthwhile exercise:

> Let me be emphatic that the uncertainty does not negate the exercise. Real footprints are the essential measure, and nothing short of them will do. The level of accuracy that I have described is good enough to separate out the flying from the hand drying.[6]

In the ten years since Berners-Lee wrote the book, a lot more research has been done and a much greater understanding gained about the importance of embodied carbon.

I have also relied heavily on the work of Hannah Richie and the Our World in Data[7] team out of Oxford University, who also build on the work of J. Poore and T. Nemecek published in 2018, who "consolidated data on the multiple environmental impacts of about 38,000 farms producing 40 different agricultural goods around the world in a meta-analysis comparing various types of food production systems."[8] In a few years, this might all be significantly easier; Unilever recently announced that it is going to calculate the footprint of all of its products, and other companies will likely follow. It might soon be as easy to measure the carbon footprint of the stuff you buy as it is to read the nutrition label on a cereal box or the Energy Star label on a new TV. Right now, it's not.

Table 2.1. The six lifestyle domains from the 1.5-degree lifestyle report

	item		source	unit	CO2/unit	estimates/day	19-Jun
	Total		daily allowance	grams	2,500,000	6,849.32	6,849
	Notes						even a 15HP "green" motor isn't very fuel efficient
Media	newspapers	daily	berners-lee		800	400.00	
	newspapers	weekend	berners-lee		1,800	900.00	
	Data	variable	JC Mortreux	gms/GB	123	62.00	
	Data	revised	finnish study and othe	gms/GB		10.00	
Transportation	e-bike		calculated	grams/km	17		
	streetcar/subway		JC Mortreux	grams/km	47		
	bus						
	Subaru			grams/km	160		1600
	Aviation						
	outboard on boat			grams/min	214.67		1288
Water heating	Shower		berners lee	each		600.00	
	Bath		berners lee	each		1,100.00	
Food	veg breakfast	est	JC mortreaux			350.00	350.00
	veg lunch	est	JC mortreaux			500.00	500.00
	veg dinner	est	JC mortreaux			500.00	500
	beef		Omni	serving		7,200.00	7200
	pork		Omni	serving		950.00	
	chicken		Omni	600		800.00	
	fish		Omni	serving		1,100.00	
	cheese		Poore	per 100g		1,100.00	
	Milk		Poore	per liter		3,200.00	
	snack					300.00	
Alcohol	red wine		food climate research	glass		275.00	
	beer		food climate research	pint		330.00	330
	martini		food climate research	double		123.00	
Basics			fixed operating (gas, electric, water)				105
			fixed embodied carbon (apple)				
							11,873
							-5,024
Basics	gas heating	fixed	calculated			1,100.00	1.73
	electricity	fixed	calculated	grams/kwh	24	105.00	
						0	
embodied	iphone	embodied	apple	kg/LCA	80	73.06	
	ipad	embodied	apple	kg/LCA	119	108.68	
	macbook	embodied	apple	kg/LCA	174	158.90	
	watch	embodied	apple	kg/LCA	40	36.53	
	Subtotal Apple products					377.17	
	TV	embodied	diane saxe	kg/year	250	684.93	
Total repeated						1,205.00	

It may not be perfect, but using the data I could gather, I started building a spreadsheet based on the six lifestyle domains from the 1.5-degree lifestyle report. With food, I quickly found that breaking up every single meal into its components was onerous and not worth the trouble. A meal without meat or dairy almost always came out about the same, so I took an approximate number and would add a factor for different foods that had a dramatic impact on the footprint. For housing, it was almost impossible to separate my own impact from the rest of the family, so I calculated a general operating cost for the house and would add baths and showers, because they have an impact I can measure independently. If there was anything that really stood out (like a takeout Chinese food dinner), I would put it in notes.

As the project evolved, the accuracy improved as I got more information. New tools and resources continue to show up, and the data continue to be revised. There is more detail about what one can learn from the data in each section, each lifestyle domain that follows.

–3–

Why Individual Actions Matter

In a candidates' debate in 2019, the question of regulating light bulbs and straws came up. Elisabeth Warren responded:

> Oh, come on, give me a break. This is exactly what the fossil fuel industry wants us to talk about.... They want to be able to stir up a lot of controversy around your light bulbs, around your straws, and around your cheeseburgers. When 70% of the pollution, of the carbon that we're throwing into the air, comes from three industries.[1]

The 70% number has been around since about 2017, when the Carbon Majors Report[2] concluded that just 100 companies in the fossil fuel industry were responsible for 71% of the world's greenhouse gas emissions since 1988. Called "a new and powerful perspective on climate accountability," the Report "offers insight into responsibility from the perspective of the producers of hydrocarbons; those companies that have made astonishing returns over decades through the extraction and production of greenhouse gas emitting products." According to the *New York Times*, "the three industries contributing to the most carbon dioxide emissions in the United States right now, Ms. Warren noted, are the building industry, the electric power industry and the oil industry."[3]

It seems that everyone throws around the 70% and the 100 companies thing, with most people quoting the *Guardian* article covering it. Tess Riley of the *Guardian* writes that "ExxonMobil, Shell, BP and Chevron are identified as among the highest emitting investor-owned companies since 1988," and when listing the top 100 producers, notes only their "cumulative greenhouse gas emissions."[4] I wonder if most people quoting the report ever actually read it, because it paints a very different picture. Only Exxon and Shell are in the top 10, with Exxon (at number 5) coming in at 2% cumulative carbon. Number 1 is a national entity called China (coal) at 14%. National entities from Saudi Arabia, Russia, and Iran are next, totalling 10.7%. So right off the bat, it is incorrect to say 100 companies; these are national governments and the entities they control. It's hard to blame BP for everything when it is at most 1.5% of the problem.

The list is out of date too: Murray Coal is bankrupt, and Peabody Energy is circling the drain. This is what happens when nobody is buying what you are selling. An analyst noted: The industry continues to be battered by rapid structural decline driven by low gas prices, the low and falling cost of building wind and solar power generation and sweeping initiatives by utilities and corporations to cut emissions.[5]

Exxon is not such a force to be reckoned with anymore either; it was just kicked out of the S&P 500 because, as an analyst put it, "oil has shrunk as part of every economy, not only the US. This is a global trend."[6]

But the biggest problem with the whole 71% complaint is that it is counting cumulative emissions, combining what are called *Scope 1 and Scope 2 Emissions* that come from actually producing the product and "arise from the self-consumption of fuel, flaring, and venting or fugitive releases of methane," and *Scope 3 Emissions* that "account for 90% of total company emissions and result from the downstream combustion of coal, oil, and gas for energy purposes."[7]

Scope 3 is us consuming, putting their gas into our cars and planes and heating our houses; it's companies making steel and aluminum and concrete that goes into our houses and cars and buildings. They are lifestyle emissions, coming from the choices we make, the things that we buy, the governments we elect.

This doesn't let the 100 entities off the hook; they have done everything they can to get us to consume more of their product. But it wasn't really hard. Nobody buys a cup of oil or electricity; they buy what it does. And it does so much, having enabled the biggest economic boom since humans first walked. The modern world was built since the Second Industrial Revolution that started in about 1870, powered by oil and electricity that is used to make the stuff that we're buying or using. The industries managed to convince us to consume, to always sell us more; to live in crappy houses and work in crappy buildings that use too much energy in places we have to use oil-consuming cars to get to. Vance Packard wrote in his 1960 book, *The Waste Makers*:

> They must learn to consume more and more or, they are warned, their magnificent economic machine may turn and devour them. They must be induced to step up their individual consumption higher and higher, whether they have any pressing need for the goods or not. Their ever-expanding economy demands it.[8]

For industry to produce, we have to consume ever-increasing amounts of fuel to power our lives and the stuff that we buy. But we saw what happens when we stopped consuming during the pandemic: Oil companies lost billions. Major frackers like Chesapeake Energy went bust. Airlines failed or laid off tens of thousands of employees. The fuel-thirsty 747 was retired. The world changed when we stopped buying what they were selling.

It changed not because of *production* but because of *consumption*, which is what really drives the economy. We stopped *buying* what they were *selling*. We didn't have much control over it, it was not a voluntary reduction, but we got a demonstration in real time of the importance of our actions and why our individual carbon footprints can make a difference.

Others are still not convinced that carbon footprinting is anything other than a corporate plot. My colleague at *Treehugger*, Sami Grover, wrote a few years ago:

> This is actually why oil companies and fossil fuel interests are all too happy to talk about climate change—as long as the focus remains on individual responsibility, not collective action. Even the very notion of "personal carbon footprinting"—meaning an effort to accurately quantify the emissions we create when we drive our cars or power our homes—was first

> popularized by none other than oil giant BP, who launched one of the first personal carbon footprint calculators as part of their "Beyond Petroleum" rebranding effort in the mid-2000s.[9]

The climate scientist Michael Mann said much the same thing in *Time*, noting that "there is a long history of industry-funded 'deflection campaigns' aimed to divert attention from big polluters and place the burden on individuals."[10]

He raises the valid point that many of these campaigns for individual actions are organized by big business, which is certainly true; the best example is the obsession with recycling, which I have described as "a fraud, a sham, a scam perpetrated by big business on the citizens and municipalities of America.... Recycling is simply the transfer of producer responsibility for what they produce to the taxpayer who has to pick it up and take it away."[11] Not only have the industries that have thrived on the linear take-make-waste convinced us to pick up their garbage, but a recent survey found that 79.9% of people around the world are convinced that it's actually the most important thing we can do for our planet.[12]

Recycling solved a big problem for industry; like the earlier "Don't be a litterbug" campaigns, it shifted responsibility from the producer to the consumer. Carbon footprinting is thought by some to be similar, especially when you see BP trying to make us feel responsible for our fossil fuel consumption instead of blaming them.

But BP didn't invent the carbon footprint; it was one of a few footprints that were part of the "ecological footprint" developed by William Rees of the University of British Columbia and Mathis Wackernagel. BP just co-opted it, and that is not a reason to throw the baby out with the bathwater. I believe it is just as dangerous and counterproductive to suggest that individual actions don't matter very much, as Michael Mann does:

> Individual action is important and something we should all champion. But appearing to force Americans to give up meat, or travel, or other things central to the lifestyle they've chosen to live is politically dangerous: it plays right into the hands of climate-change deniers whose strategy tends to be to portray climate champions as freedom-hating totalitarians.[13]

If we are worried about playing into the hands of climate-change deniers, then we have already lost. They already think we hate their freedoms; as Sebastian Gorka, former Deputy Assistant to Donald Trump, said about the Green New Deal: "They want to take your pickup truck. They want to rebuild your home. They want to take away your hamburgers." It's true; we do. However, it is not likely to happen in our current political system, and that doesn't mean I *have* to drive an F150 to McDonalds.

Mann instead calls for "political change at every level, from local leaders to federal legislators all the way up to the President." I agree, but anyone who watched the last American election knows how that worked out—they may have changed the President, but the party of climate deniers and delayers actually increased their control everywhere else. Furthermore, this whole discussion is setting up another diversion, another division. Do we just eat our burgers, drive our pickup truck, and say I'm waiting for system change? Or do we try to set an example?

As Leor Hackel and Gregg Sparkman suggest in a *Slate* article titled "Reducing Your Carbon Footprint Still Matters":

> Ask yourself: Do you believe politicians and businesses will act as urgently as they need to if we keep living our lives as though climate change were not happening? Individual acts of conservation—alongside intense political engagement—are what signal an emergency to those around us, which will set larger changes in motion.[14]

Of course, it requires more than individual action; it requires political action, regulation, and education. Perhaps the best example is the campaign against smoking, where we saw what happens when individuals, organizations, and government work together. Smoking was promoted by the industry, who buried information about its safety and owned the politicians and fought every change. They hired experts and even doctors to challenge the evidence and deny that smoking was harmful. They had a real advantage in that the product they were selling was physically addictive. However, eventually, in the face of all the evidence, the world changed.

Forty years ago, almost everyone smoked, it was socially acceptable, and it happened everywhere. Governments applied education, regulation, and taxes. There was a lot of social shaming and stigmatizing happening too; in 1988, medical historian Allan Brandt wrote, "An emblem of attraction has become repulsive; a mark of sociability has become deviant; a public behavior now is virtually private."[15] Instead of virtue-signalling, we had vice-signalling.

But this shift also took a great deal of individual determination and sacrifice. You can talk to almost anyone who was addicted and has given up smoking, and they will tell you that it was the hardest thing they have ever done.

Fossil fuels are the new cigarettes. Their consumption has become a social marker; look at the role pickup trucks played in the 2020 American election. Like cigarettes, it is the secondhand externalized effects that are the motivators for action; people cared less when smokers were just killing themselves than they did when secondhand smoke became an issue. I wonder if at some point the big obnoxious pickup truck won't be as rare as smokers have become.

It's Not Just Virtue Signalling

In writing about my carbon footprint, I have been accused of "virtue-signalling," described by *Financial Times* writer Robert Shrimsley as "the apparently modern crime of trying to be seen doing the right thing."[16] He noted criticism of actor Emma Thompson, who showed up at an Extinction Rebellion one day and was seen flying first class the next:

> She is one of life's enthusiasts and throws herself into supporting any number of liberal causes without first checking off every aspect of her own behaviour. She has never denied that she flies, but once the virtue-signalling line of attack kicks in, you can never do anything right. Fly and you are a hypocrite, turn your home into a sanctuary for displaced orangutans and you are just a virtue signaller. It's a curious outlook—must all campaigners be fanatics?[17]

No, nor are they freedom-hating totalitarians. They are just people trying to do the right thing.

I will be out there demanding political action and voting for the parties and politicians with the strongest climate platforms. When they get elected, I harangue them individually and in my writing. I support organizations financially, have sat on boards, and have protested and marched. I will continue doing that, but in the end, I believe that individual action matters as much or more than political action because it happens every day, every minute. I vote every four years, but I eat three times a day.

Production vs Consumption and the IPCC

Production-based accounting (PBA) measures emissions from everything that happens within a nation's borders; it is the basis of the Paris Agreement, in which the countries that signed agreed to reduce their own emissions. However, over the past few decades, thanks to globalization, there has been significant de-industrialization in developed countries and a shift of production to developing countries; our carbon emissions have been offshored. That's why consumption-based accounting (CBA) of carbon makes so much more sense. As one study noted:

> PBA provides an incomplete picture of driving forces behind these emission changes and impact of global trade on emissions, simply by neglecting the environmental impacts of consumption. To remedy this problem, several studies propose to consider national emissions calculated by consumption-based accounting (CBA) systems in greenhouse gas (GHG) assessments for progress and comparisons among the countries.[18]

If I buy a Haier air conditioner or a Samsung washing machine, who is responsible for the emissions that come from making that machine? Should I be able to offshore the emissions involved in their manufacture when it is my choice to buy the machine? We blame China for all that pollution from their factories, but they are making the stuff that we want and we buy. That's why it's better to measure *consumption*: it is a more accurate measure of what each party, whether an individual or a nation, is responsible for. But to do that, we have to determine how much carbon was emitted while making that air conditioner or washing machine.

The Importance of Upfront Carbon Emissions and Embodied Carbon

Measuring these emissions creates a problem, because it's not easy or simple. The operating energy and carbon, for example, the fuel for your car or the electricity for your air conditioner, are relatively straightforward but you also have to measure what's commonly called the *embodied carbon* or *embodied energy*. I have always thought this was a terrible name because the CO_2 or carbon is not embodied in what you are consuming, it is in the atmosphere already. These are the emissions that come from making the steel, aluminum, and plastic in your Chinese Haier air conditioner; bringing it all together and manufacturing it; shipping it to North America; and then getting it to your home. These are what I call *upfront carbon emissions*, or UCEs. Others are coming to agree with me; the World Green Building Council has adopted the term.[19]

It's very easy to read that Energy Star label on the air conditioner to see how much energy it uses, how many kilowatt-hours of electricity, the operating energy, and from that you can simply calculate the carbon footprint of operating it. Upfront carbon is another story; there are so many variables. The *Economist* gives an example:

> A medium-sized electric-car battery made in Sweden, which uses lots of renewable energy, emits 350 kg of carbon dioxide. The same battery made in Poland, which relies on coal, emits over 8,000 kg. The mode of transport matters, too—goods that are transported by aircraft are far dirtier than those carried on ships.[20]

Some industries are starting to take embodied carbon seriously; the building industry is looking at using more wood. But notwithstanding the success of the DeHavilland Mosquito or the Morgan sports car, it's tough to build an airplane or a car out of wood.

The *Economist* concludes that the world needs to shift toward goods that have a cleaner footprint, regardless of where they are produced and that "if climate change is to be tackled, countries and consumers must take full responsibility for their carbon."[21]

When you start thinking seriously about embodied carbon as well as operating carbon, it changes everything.

C40 Cities: The Future of Urban Consumption in a 1.5-Degree World

The bias toward looking at production rather than consumption has long distorted the thinking about sustainability. There have been many books like David Owen's *The Green Metropolis* that suggested cities like New York are the greenest places to live because the carbon footprint per capita is so low. However, the C40 Cities organization, "a network of the world's megacities committed to addressing climate change," recently produced a report with ARUP and the University of Leeds that looks at the picture from a consumption point of view.

> Take a pair of jeans, for example. Its climate impact includes the GHG emissions that resulted from growing and harvesting the cotton used for the fabric, the CO_2e emitted by the factory where it was stitched together, and the emissions from ships, trucks or planes that transported it to the store. Its impact also includes the emissions from heating, cooling or lighting the store the jeans were bought in and the CO_2e emitted by the end-consumer washing and drying it over its lifetime.[22]

Much of what we consume is like that pair of jeans. C40 estimates that "85% of the emissions associated with goods and services consumed in C40 cities are generated outside the city; 60% in their own country and 25% from abroad."[23] The C40 report has its own list of consumption categories that differ from the 1.5-degree lifestyle report, and the emissions from each:

- Buildings and infrastructure: 11% of total emissions in C40 cities in 2017
- Food: 13%
- Private transport: 8%
- Clothing and textiles: 4%
- Electronics and appliances: 3%
- Aviation: 2%[24]

It seemed an odd set of categories and an odder distribution of emissions, but the C40 cities are big so there is a lot of public transportation. It was surprising to see clothing and textiles having twice the emissions of aviation, but everybody has to eat and get dressed but only a very small proportion of the population flies regularly.

The C40 report is more prescriptive and sometimes silly (you can only buy three articles of clothing per year! Keep your computer for 7 years! You may fly only one short-haul flight every three years!) Their proposed consumption interventions only cover 40% of urban emissions; the rest are assumed to be out of individual control. However, it confirms the importance of measuring our consumption and provides some useful data; I never knew clothing and textiles were such a big deal.

1.5-Degree Lifestyle Is Easier for Some than for Others

Years ago, the environmental thinker Alex Steffen wrote a brilliant article titled "My Other Car Is a Bright Green City"[25] that profoundly influenced me. Writing before Teslas were even on the road, he noted that "the answer to the problem of the American car is not under the hood." He continued: "There is a direct relationship between the kinds of places we live, the transportation choices we have, and how much we drive. The best car-related innovation we have is not to improve the car but eliminate the need to drive it everywhere we go."[26]

For at least a decade, I have been arguing that Steffen had things backwards. His chapter title was "What We Build Dictates How We Get Around." I argued that how we get around determines what we build, that it was the transportation technology that determined our land use patterns and built forms. But lately I have come to realize that we were both wrong; as transportation consultant Jarrett Walker tweeted last year, "Land use and transportation are the same thing described in different languages." It is not a chicken-and-egg, a which came first thing. It is a single entity or system that has evolved and expanded over the years through the changes in the form of energy available, and in particular the ever-increasing availability and reduction in the cost of fossil fuels.

This can make lifestyle changes extremely difficult for some, and easier for others. For example, I live in a streetcar suburb, where an electrified line was pushed west over a ravine in 1913 to farmland and was almost completely built out in about 20 years. The main street with the streetcar line had shops and groceries and everything you need. The houses were all built on narrow lots and relatively high density because everyone wanted to be able to walk to the streetcar in under 20 minutes. The entire development pattern was predicated on the fact that people didn't have cars, but could work downtown, get a streetcar home, pick up what they needed for dinner, and walk to their tidy house. The stores have changed over the last century, but not much; I can still patronize the butcher and the baker. The A&P where my mom shopped 60 years ago is a gym now, but there is a bigger grocery where the racetrack used to be, and a medical clinic has opened. The old movie theatre has turned into a gym, and the synagogue where I went to nursery school is now a condo site (I have lived in the neighborhood a long time!), but really, I can find almost anything I need right in my neighborhood, and if I can't, it is a streetcar or bike ride away. Because the entire pattern of development surrounding me, the world in which I live, was designed 120 years ago in a world when people didn't drive.

My family was never poor, and I was able (with a little help from the BOMAD, the Bank of Mom and Dad) to buy my first house at the age of 30 for a shockingly low price, open an architectural practice, and be pretty much self-employed from that time on. I have to acknowledge upfront that I am a very lucky boomer.

I need to say this because I always have to remember that it's relatively easy for me to live a 1.5-degree lifestyle; I live in a place where I don't have to drive and can walk to the fancy healthy butcher and organic grocer. I work at an internet-based job where I don't have to go to a factory or an office downtown; I can just go downstairs to the home office that I designed. And I can't write this book looking through my rose-colored glasses because it has to work for everyone.

I just happen to be lucky enough to live in a land use/transportation nexus that was designed 110 years ago at the dawn of the electric age, and most North Americans now live places designed in the gasoline age. To expand on Jarett Walker, "Land use, transportation, and energy are the same thing described in different languages."

And we are all stuck in an economic structure that, by design, encourages or in fact forces us to consume more energy, in particular fossil fuels, which drives the economic growth that is raising millions out of poverty and maintaining our Western lifestyles because, according to author and professor Vaclav Smil, energy and the economy are also the same thing described in different languages.

-4-

Energy, Efficiency, and Sufficiency

According to Vaclav Smil, in his book *Energy and Civilization: A History*, energy is money, the universal currency. Energy drives everything, and the more we have of it, the cheaper it is, the more the economy booms.

> To talk about energy and the economy is a tautology: every economic activity is fundamentally nothing but a conversion of one kind of energy to another, and monies are just a convenient (and often rather unrepresentative) proxy for valuing the energy flows.[1]

Our economies grew as we moved to more concentrated forms of energy; from burning wood, then coal for steam engines, which ran generators giving us electricity, which ran motors that moved streetcars, changed industry and architecture. Gasoline packed in more density and was easily transportable, giving us cars, trucks, and tractors. Natural gas made artificial fertilizers that replaced manure and allowed an explosion of food production and, with it, the population. Smil writes:

> All of these developments have combined to produce long periods of high rates of economic growth that have created a great deal of real affluence, raised the average quality of life for most of the world's population, and eventually produced new, high-energy service economies.[2]

In his more recent book, *Growth*,[3] Smil is even more direct, quoting economist Robert Ayres, who wrote: "The economic system is essentially a system for extracting, processing and transforming energy as resources into energy embodied in products and services." *The purpose of the economy is to turn energy into stuff.*

If that is what the economic system does, how do you turn down the dial from a high-energy economy that has done so much? Smil is not convinced that a transition to a low-energy economy is likely, but instead calls for a "delinking of social status from material consumption." Do we really need so much?

> Modern societies have carried this quest for variety, leisure pastimes, ostentatious consumption, and differentiation through ownership and variety to ludicrous levels and have done so on an unprecedented scale...Do we really need a piece of ephemeral junk made in China delivered within a few hours after an order was placed on a computer? And (coming soon) by a drone, no less![4]

Conscious Decoupling

Smil calls for *delinking*. Another term that is used a lot when talking about climate and the economy is *decoupling*, the idea that the economy can actually be separated or decoupled from carbon emissions. Caroline Kormann interviews optimists who claim:

> "It's absolutely the case that emissions and growth can be decoupled."...."The technology is available to have faster economic growth while reducing over-all emissions." But the switch to nuclear and renewables needs to happen more rapidly. "It takes policy. It won't happen through markets alone."[5]

One only has to look at policies of governments all over the world to see that decoupling, if it is happening at all, is simply not happening at a rate that makes any difference. Neither American political party has any intention of giving up on fossil fuels, and Canada does everything it can to keep the oil sands going in the face of all economic or climate sensibility.

Smil is dismissive of those who believe decoupling will happen, writing in *Growth*:

> Of course, most economists have a ready answer as they see no after-growth stage: human ingenuity will keep on driving eco-

> nomic growth forever, solving challenges that may seem insurmountable today, especially as the techno-optimists firmly anticipate wealth creation progressively decoupling from additional demand for energy and materials.[6]

In his book *Less Is More: How Degrowth Will Save the World*, Jason Hickel doesn't think much of decoupling either:

> Some people try to reconcile this tension by leaning on the hope that technology will save us—that innovation will make growth "green." Efficiency improvements will enable us to "decouple" GDP from ecological impact so we can continue growing the global economy for ever without having to change anything about capitalism. And if this doesn't work, we can always rely on giant geo-engineering schemes to rescue us in a pinch. It's a comforting fantasy. In fact, I once believed it myself.[7]

When Gwyneth Paltrow split with her husband, she described it as "conscious uncoupling," to much derision. In fact, it is a five-step process developed by psychologist Katherine Woodward Thomas that involves "reclaiming your power and your life" and "generating a positive future."[8]

I am going to introduce a new term here, *conscious decoupling*, making decisions in our personal lives to separate, to decouple, the activities that we do and the things that we buy from the fossil fuels that are used to run them or make them, without giving up nice things. (I like nice things.) The idea is that one can still live a nice life where there actually is growth, development, improvement, satisfaction, and a positive future without running on gasoline. A life where we can reclaim our power and generate a positive future.

The most obvious example is consciously decoupling our personal transportation from fossil fuels by riding a bike or walking instead of driving. We could consciously decouple our lawns from fertilizers by replacing them with vegetable gardens. We could consciously decouple our diets by eating locally. I decoupled my winter sports by switching from driving two hours to snowboard at a ski hill to cross-country skiing in a neighborhood park. None of these changes caused any kind of dramatic decline in my quality of life, but they did significantly decouple what I did from fossil fuels and carbon, both embodied and operating. We can all make choices that consciously decouple our consumption from our carbon footprint; it is the whole point of this exercise. Or is this just a comforting fantasy?

Is Degrowth the Answer?

Many, like Smil and Hickel, do not believe that we can have decoupling without degrowth, defined by Samuel Alexander as "a phase of planned and equitable economic contraction in the richest nations, eventually reaching a steady state that operates within Earth's biophysical limits."[9] Hickel blames capitalism and inequality.

> The data on this is clear: people who live in highly unequal societies are more likely to shop for luxury brands than people who live in more equal societies. We keep buying more stuff in order to feel better about ourselves, but it never works because the benchmark against which we measure the good life is pushed perpetually out of reach by the rich (and, these days, by social media influencers). We find ourselves spinning in place on an exhausting treadmill of needless over-consumption.[10]

I was initially skeptical about degrowth as discussed by Hickel, thinking it romantic and unrealistic, until I realized that I have been talking about it for years under a different name: *sufficiency*. Alexander described it earlier as the sufficiency economy:

> This would be a way of life based on modest material and energy needs but nevertheless rich in other dimensions—a life of frugal abundance. It is about creating an economy based on sufficiency, knowing how much is enough to live well, and discovering that enough is plenty.[11]

Sufficiency is the key concept, the thread that runs through this book. It doesn't mean the end of growth; people still need a roof over their head, a way to get around, food to eat, and clothing to wear. We still need to grow a housing industry that can build efficient homes out of natural materials. We still need to grow a food industry that doesn't rely on natural gas. We still need a transportation industry that builds electric trains and e-bikes. And of course, we all still need those services like health care and medicine that are such a big part of our economies now. But we can do it with a lot less stuff per person. Alexander succinctly explains why sufficiency is so much more important than efficiency:

> Everyone knows that we could produce and consume more efficiently than we do today. The problem is that *efficiency without sufficiency is lost* [emphasis added]. Despite decades of extraordinary technological advancement and huge efficiency improvements, the energy and resource demands of the global economy are still increasing. This is because within a growth-orientated economy, efficiency gains tend to be reinvested in more consumption and more growth, rather than in reducing impact.[12]

The reason efficiency without sufficiency is lost goes back to Stanley Jevons, who in 1865 noted that when James Watt developed a steam engine that burned 75% less coal than the Newcomen engine it replaced, consumption of coal did not go down but in fact went up dramatically, as people figured out how to use that steam engine in so many different ways. It is now called Jevons paradox or the rebound effect: as engines get more efficient, we get SUVs; as insulation gets better, houses get bigger. As airplanes get more efficient, more people fly, so the energy and carbon savings are not as big as they might be. The rebound effect is often used as an excuse to not do anything to increase efficiency and is beloved of climate skeptics. Others claim that it is over-rated and note that just because a car is twice as efficient, people don't drive twice as far. However, many are buying cars that are twice as big with larger upfront carbon emissions and more carbon spewed for every road mile.

Also, the Watt steam engine wasn't just an incremental improvement, it was a technological revolution that changed how steam engines were used; they were now efficient enough that they could power boats and trains and factories rather than just pumping water. The range of uses exploded and, with it, coal usage. It was very different than what we are seeing with incremental improvements in technologies like cars or airplanes.

But we are seeing this with LED lights. They are not just a vastly more efficient light source but are also being used in different ways. I have seen building interiors, people's living rooms where you could do surgery, there is so much light. Buildings are encrusted with them. LED billboards that run all day. TVs are now the size of movie screens. Measurements from space show massive increases in the amount of light being emitted. Many people are getting smart bulbs that they control with their phones; I have them over my dining room table. These are constantly talking to the controller, which then talks to the internet and my phone. I have calculated that they use more energy while they are off, waiting for instructions, than they do for the one hour we might have dinner under them. I have argued (but have not been able to prove) that it is likely that as the uses of LEDs explode, their power consumption might well soon add up to as much as the lighting it replaced. That is Jevons paradox in action, eating up much of the gains in efficiency.

This is just one example of how even dramatic increases in efficiency can eat themselves, and why the concept of sufficiency is so important.

The Rich Are Different Than You and Me

It would be nice if the rich actually used their money to reduce their energy consumption. While they have the means and ability to live a low-carbon lifestyle and are the primary consumers of Passive House dwellings, Tesla cars, heat pump dryers, and photovoltaic roofs, not to mention happy meat and organic produce, the vast majority of them don't think about doing so. And even the Tesla-driving, grass-fed meat-eating professed environmentalist flies, a lot. A recent European study found that "the EU top 1% emit 55 tCO_2 eq/cap on average, more than 22 times the 2.5-tonne target."[13] Of that, 41% is from flying. Among the EU top 10%, almost a third of their emissions are from travel. My flying in 2019 was responsible for 6.35 tonnes of CO_2 emissions, and that was historically low for me, only two transatlantic trips. Another article, unsubtly named "Scientists' Warning on Affluence," claims that the world's richest 10% emit 43% of the carbon and totally blames the affluent society for our problems:

> The affluent citizens of the world are responsible for most environmental impacts and are central to any future prospect of retreating to safer environmental conditions. Any transition

> towards sustainability can only be effective if far-reaching lifestyle changes complement technological advancements.[14]

The problem is that, in global terms, billions of people have been lifted out of poverty, the world has been getting richer, and when people get money, they buy stuff. They travel. Consumption is a direct result of affluence, and CO_2 is a direct result of consumption.

The study authors conclude that we have to change the way we live, the way we think about wealth, calling for "the adoption of less affluent, simpler and sufficiency-oriented lifestyles to address overconsumption—consuming better but less."

It all comes back to Smil: the need for "delinking of social status from material consumption."

Radical Efficiency vs Radical Sufficiency

The affluence study has an interesting graph that shows how the world is getting more efficient; the latest jets consume 20% less fuel per passenger mile, and the energy intensity, or amount used to make things, is going down. However, the global gross domestic product and the global material footprint are still going up fast. CO_2 emissions are going up too, albeit at a slower rate. But that's not good enough; CO_2 emissions have to go down. The only way to do that is through *sufficiency*, a term I learned from Kris De Decker, who wrote:

> The problem with energy efficiency policies, then, is that they are very effective in reproducing and stabilising essentially unsustainable concepts of service. Measuring the energy efficiency of cars and tumble driers, but not of bicycles and clotheslines, makes fast but energy-intensive ways of travel or clothes drying non-negotiable, and marginalises much more sustainable alternatives.[15]

The clothes dryer is one of my favorite examples. The standard tumble dryer is incredibly inefficient, taking conditioned air from inside the home, heating it with toaster coils, evaporating moisture out of the clothing, and then dumping all that heat and moisture outside. You then have to pay again to heat or cool the air that comes into the home to replace the air that was exhausted. In highly energy-efficient homes, this can be a major issue, so now you can pay significantly more to get a heat pump dryer that cleverly dries the clothes and evaporates the moisture, then takes the heat out of the exhaust and condenses the water, creating a closed loop. It saves a lot of energy and eliminates the exhausted air but is expensive and complicated. Meanwhile, in many parts of the world (including some that are cold), people do not like dryers and insist that air-drying on a line is the only way to treat clothing properly. In China you will see some of the wealthiest homeowners still putting clothes out on the line.

In a world focused on sufficiency rather than efficiency, you build a better clothesline. Sufficiency is a key concept for getting out of this mess. One has to ask about everything that we do and buy: how much do we need? What is sufficient for the job? Can I walk or ride a bike instead of driving? Do I need such a big house, a big car, or my personal bête noir, do I need that new Apple product? Sufficiency becomes particularly relevant when you consider embodied carbon, because buying anything, even the most energy-saving wonderful new dryer or electric car or computer's footprint, affects us right now.

A recent research report from Finland, *The Sufficiency Perspective in Climate Policy: How to Recompose Consumption* (Why are all these great studies from Finland?) examined a number of issues we have been looking at, including consumption vs production and efficiency vs sufficiency.[16] The report noted that "despite remarkable increases in efficiency, total emissions have not decreased." It calls for sufficiency, the downscaling of consumption:

> Sufficiency can have different forms: reduction and consuming less includes examples such as driving fewer kilometers or eating less meat. Substitution and fulfilling needs in another way means for example shifting to public transport from a private car, to a plant-based diet from a diet with a lot of meat or partly replacing clothes washing by airing clothes. Adjusting consumption to meet needs can include lowering room temperatures and reducing apartment size in relation to the number of inhabitants.[17]

But where most arguments for sufficiency are about personal choices, the study notes that it can actually be promoted through government policy, "to create conditions that enable sufficient practices and support sufficient lifestyles to become routinised."

Samuel Alexander of the Simplicity Collective (sufficiency and simplicity keep getting mixed up together) dismisses "green growth" and describes a "sufficiency economy":

> It is of the utmost importance, of course, that we use the best of our technological knowledge to help us achieve a sustainable way of life through efficiency improvements. It would be foolish to argue otherwise. But efficiency alone cannot "decouple" economic growth from ecological impact sufficiently to produce a sustainable way of life. The extent of decoupling required is simply too great. To be effective, the drive for efficiency must be shaped and limited by an ethics of sufficiency. That is to say, our aim should not be to do "more with less" (which is the flawed paradigm of green growth), but to do "enough with less" (which is the paradigm of sufficiency).[18]

"Enough with less" is an interesting concept. Take or build as much as you really need, and make it with as little stuff as possible, to run as efficiently as possible. As the great designer Dieter Rams noted in his ten rules: *Good design is as little design as possible*: "Less, but better—because it concentrates on the essential aspects, and the products are not burdened with non-essentials. Back to purity, back to simplicity."[19]

– 5 –

What We Eat

From the 1.5-degree lifestyle report: Nutrition—"intake of all foodstuffs and beverages consumed at home and outside the home, e.g., vegetable and fruit, meat, fish, dairy, cereal, alcohol and non-alcoholic beverages. Direct emissions from cooking at home are included under housing, whereas emissions from operation of restaurants are included under leisure."

How We Measure

Here is an example of how hard this process is. My family has always loved the rotisserie chicken from the Canadian Swiss Chalet chain, but we stopped ordering it because of the appalling amount of plastic waste that came with it, mainly from these custom-molded plastic shells that enclose each order. However, in the spirit of this project, we ordered it for a Saturday night dinner recently so that I could look at the various components of its carbon footprint.

Dinner: Swiss Chalet rotisserie chicken, ¼ chicken, white meat, ordered for delivery.

- Raising chicken: 1,431 grams
- Cooking chicken: 72 grams
- Packaging chicken: 366 grams
- Delivering chicken: 2,737 grams
- Total: 4,606 grams

Raising the Chicken

Most of the data in this book on the carbon footprint of food come from Our World in Data,[1] a project based in Oxford University that is "about research and data to make progress against the world's largest problems." They break it up into components including land use (2.5 kg CO_2 per kg for chicken), farm (fertilizers etc., 0.7), animal feed (1.8), processing (0.4), transport (0.3), retail, and packaging. Since this chicken is not going to the store and is packaged differently, I deleted the last two categories.

But are numbers generated at Oxford University relevant to a chicken in Toronto, Canada? Canadian chickens are probably raised in cages at higher density than in the UK, where standards for animal welfare are tougher. But transport may be far higher because of greater distances traveled, and globally it consumes a huge amount of energy; this is the "cold chain," which "involves the transportation of temperature-sensitive products along a supply chain through thermal and refrigerated packaging methods and the logistical planning to protect the integrity of these shipments" and which "could use up as much as 15 percent of the world's energy production."[2]

But these are the only numbers I have, so I multiply the total by the weight of my chicken portion, realizing as I write this that some of the weight of the chicken is lost during cooking as the fat drains off.

Cooking the Chicken

I searched for commercial gas-fired rotisserie ovens and found they use 55,000 BTUs of gas per hour. I divide that by the number of chickens it holds and how long they cook, and then again by four since I am eating a quarter chicken, coming up with 72 grams.

But why are we stopping there? What about the rest of the kitchen? It has a fancy ventilation system, heating and cooling, an exhaust hood over the fryers where they make the French fries. (I haven't done the math on the fries because the carbon footprint of root vegetables like potatoes is negligible, and I have no way of knowing where they come from and how far they have travelled.)

Delivering the Chicken

This should be really straightforward, right? Just look at what kind of car the delivery guy drives, multiply its mileage rating by the distance to figure out fuel consumption, then convert litres of gasoline to CO_2. Bingo: a shocking 2,737 grams, by far the biggest item on the list so far.

But there are so many judgements here. There is a Swiss Chalet restaurant 3 km from my home, but the company has chosen to fill orders from one 7 km away. Most significantly, I ordered dinner for four people, but have attributed all of the CO_2 just to my dinner, because I *could* have ordered for one.

Then there is the question of whether fuel consumption is the only thing that should be measured. I go on in this book about the importance of measuring embodied carbon, the upfront emissions from making something like the driver's Toyota Corolla. It has an embodied carbon load of about 10 tonnes; divide that by the estimated life of the car (300,000 km) and you get 33 grams of carbon per kilometer, so his drive to my house and back would add 466 grams of embodied carbon to the trip. I didn't include that because you can only go so far down this rabbit hole before you realize that it is endless; our carbon footprint follows so many paths, has so many components, so many variables that it is really almost impossible to capture it all.

Packaging the Chicken

Finally, it arrives in paper bags with stickers stating "We're taking steps to reduce our plastic packaging! Like our new paper packaging? Send us your feedback!" I am thinking that I will do exactly that when I find the same old heavy plastic clamshells holding the chicken inside the bag, the ones that have made me give up on Swiss Chalet. I weigh it and multiply that by the CO_2 emitted in making plastic and find that it totals only 366 grams, a meaningful number but a fraction of the impact of growing a chicken or delivering it. All these years, I have been obsessed with the packaging, losing sight of the much bigger problems.

All of this is a very long-winded way of saying that carbon footprinting is hard. With so many variables, it is not an exact science. We can only approximate it.

What We Eat

A decade ago, everyone was talking about the 100-mile diet, an experiment carried out by Vancouver writers Alisa Smith and J. B. MacKinnon where they ate food grown within 100 miles of where they live, promoting the concept of local food. It was a fascinating and successful venture, turning into a book and a TV series.

I was writing about environmental issues for *Treehugger* and *Planet Green*, a *Discovery TV* network that showed the *100-Mile Diet* series. Meanwhile, I would go to dinner at my mom's house, and in the middle of winter, she would be serving fresh asparagus, flown in from Peru. I would explain how bad this was, and she would respond that she had grown up in the Depression, when she would have no vegetables but potatoes. Then in the fifties, she had canned vegetables, and in the sixties, frozen. She thought that it was just about the most wonderful thing in the entire world that she could go to the neighborhood supermarket in the middle of winter and get fresh asparagus, and she was proud to serve it.

Meanwhile, I was having my own doubts about the 100-mile diet, and about the importance of local food. I was concerned about carbon emissions, and low mileage didn't necessarily mean low carbon; research showed that a local hothouse tomato served out of season had a far higher carbon footprint than a tomato trucked up from Mexico.

My wife was writing about food for *Planet Green* at the time, and we became convinced that a seasonal diet was as important as a local one and ran our own experiment: the nineteenth-century Ontario diet. Ontario has a short growing season and a long winter (much longer in the nineteenth century), so the diet was wildly variable. In the spring, the asparagus was succulent and wonderful, so much better than my mom's. In July, the strawberries were to die for, bearing no resemblance to the wooden ones from California. In August, tomatoes. So many tomatoes, in so many salads and pastas. Kelly would buy them and everything else by the bushel, and spend weeks canning them to get us through the winter.

By December or January, it was a different story. Potatoes. Turnip. Parsnips. More turnip. Root vegetable after root vegetable, and meat, lots of pork and beef because, back in the nineteenth century, animals could huddle together to stay warm until they were butchered and served. There are many interesting ways to cook a potato or a turnip, but by March we were counting down the days until asparagus.

We are not so doctrinaire anymore and will buy a head of lettuce in January and a few limes for a margarita. But neither of us ever want to look at a California strawberry or a hothouse tomato again.

Figure 5.1. Greenhouse gas emissions per 1,000 kilocalories
Greenhouse gas emissions are measured in kilograms of carbon dioxide equivalents ($kgCO_2eq$) per 1,000 kilocalories. This means non-CO_2 greenhouse gases are included and weighted by their relative warming impact.
Note: Data represents the global average greenhouse gas emissions of food products based on a large meta-analysis of food production covering 38,700 commercially viable farms in 119 countries.
Source: Poore, J., & Nemecek, T. (2018). Additional calculations by Our World in Data.

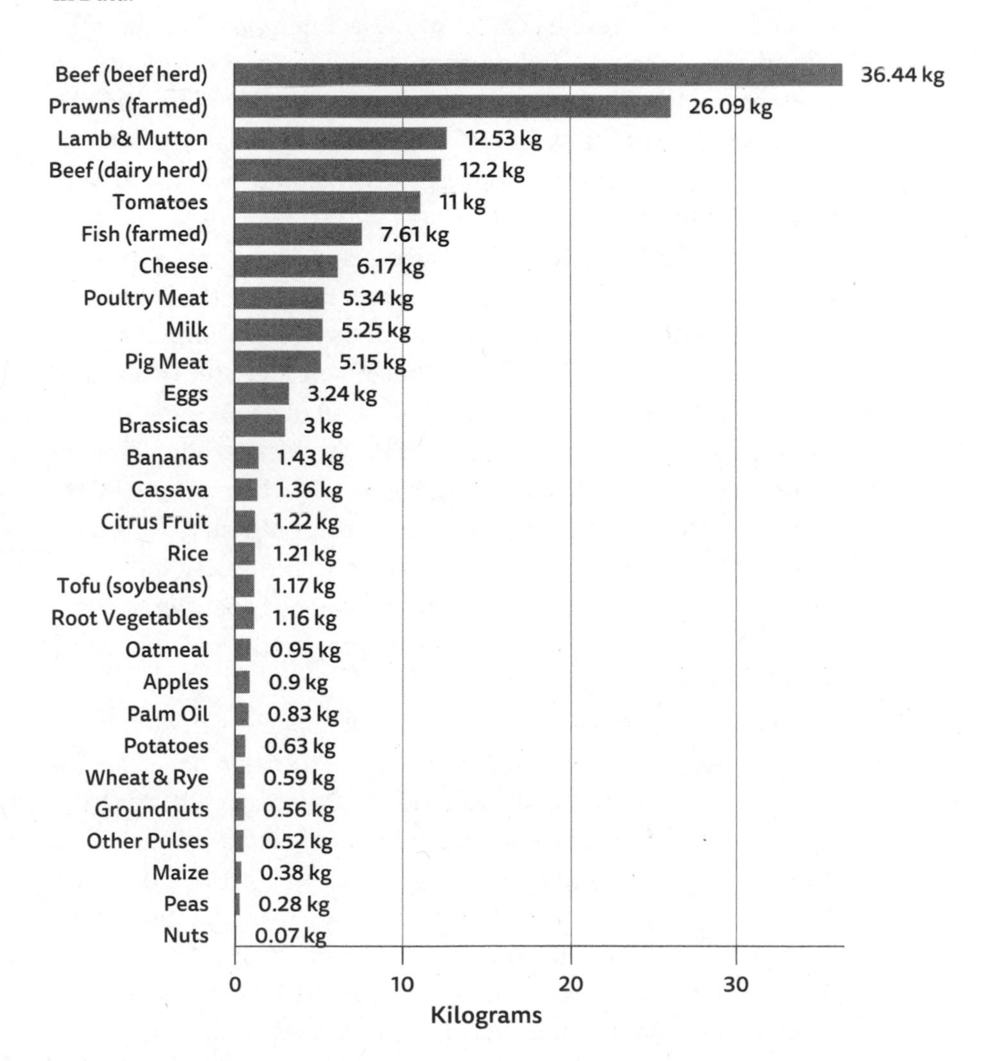

The biggest difference between today's low-carbon diet and our nineteenth-century local and seasonal diet is the protein, the meat, particularly the beef. According to Our World in Data,[3] producing 1,000 kilocalories (or dietary calories) of meat emits 36.44 kg of carbon dioxide equivalents (CO_2e).

I should explain that, for food, I prefer to use the CO_2 per 1,000 calories rather than the CO_2 per kg of food because they have such different caloric densities. A kg of lamb has a footprint of 24 kg of CO_2e, and 1 kg of cheese has 21 kg of CO_2e, but as my daughter the cheesemonger keeps reminding me, people don't sit down and eat 1 kg of cheese. When you look at it in terms of CO_2e per 1,000 calories, the lamb emits 12.53 kg and the cheese less than half that, at 6.17, sitting between chicken and fish, which is way below hothouse tomatoes at 11 kg CO_2e.

Using the calculations of greenhouse gas emissions per 1,000 calories also makes it possible to do the math. I can use My Fitness Pal or other calculators to figure out the calories of everything that I eat and can then convert them to CO_2e. It becomes easy to see that a meal based on rice or tofu has a footprint that's 1/30 of beef.

This is the other advantage of the modern low-carbon diet over the nineteenth-century Ontario diet; those English and Scottish immigrants to Canada wouldn't have touched these alternative proteins like tofu or beans that can deliver a delicious and varied diet without any meat or dairy. They didn't have cookbooks about Indian and Asian and Mexican and other cuisines where you can make so many wonderful dishes without meat.

CO_2 emissions drop like a stone if you follow a vegan diet. However, a vegetarian diet that includes eggs and dairy has probably three times the footprint and, realistically, is not much better from that point of view than one that includes pork or chicken. Vegan diets are meaningful regarding CO_2, but vegetarian ones actually don't appear to make much difference if you keep away from the ruminants.

Cut out the beef and lamb and you have a great start to a low-carbon diet. That's the *modal shift*, the change in what we eat.

But there is also the question of *sufficiency*, about how much we eat and how much we waste versus how much we actually need.

The Footprint of Food Waste

Between 40 and 50% of all food produced in North America is wasted, so every number in that carbon footprint of food can really be increased by half to account for it. According to the U.S. Department of Agriculture, there are two kinds of wasted food: *food loss*, which happens in the chain from field to fridge, whether it is left in the field, bycatch thrown overboard, losses in processing, or spoiled in transit. It's kind of the cost of doing business, and much of it is beyond our control (although again, we can choose a short food chain and reduce it). *Food waste* (about half of total waste) is more conscious: prepared food and baked goods that are thrown out, vegetables that aren't pretty, oversized portions in restaurants (on average, diners leave 17% of their meal on the plate), composting vegetables at the back of the fridge, stuff thrown out because it somehow magically changed on its best-by date. According to a study by the Natural Resources Defence Council (NRDC), about two thirds of home food waste comes from food thrown out without being used due to improper storage or "lack of visibility in refrigerators."[4] The other third comes from over-preparing, making too much food and not eating the leftovers.

This all has a number of different carbon impacts: the straightforward losses during production, the methane from food rotting in landfill sites, and the energy of preparation. According to a McKinsey study[5] quoted by FoodPrint, "household food losses are responsible for eight times the energy waste of farm-level food losses due to the energy used along the food supply chain and in preparation."[6]

How Much Do We Actually Need to Eat?

But there is another type of food waste that doesn't get discussed enough, and that is how much more food we eat than we actually need. Everything is bigger than it used to be; according to the NRDC, "The Cornell Food and Brand Lab reports that serving sizes in the *Joy of Cooking* cookbook have increased 33.2% since 1996. A recipe that used to 'serve 10' now 'serves 7' or the ingredient amounts are greater for the same number of servings."[7] According to a British report with the wonderful name *Portion Distortion*, "In 1993 an average portion of American muffins was 85 g—today, these portion sizes vary from 72 g to 130 g."[8] Everything has been supersized. Even healthy foods like bagels are 24% larger than they were 30 years ago. And as Marion Nestle wrote in her book *What to Eat*, "It is human nature to eat when presented with food, and to eat more when presented with more food."[9]

Superficially, one can say that eating portions that are a third larger means that the CO_2 emissions from making the food are a third more. But it is much worse than that because, around the world, it has led to a massive obesity crisis with a carbon footprint all of its own.

One study on the carbon footprint of obesity found that "a population with an abnormally high mean body mass index (BMI) and 40% obesity requires 19% more food energy for maintenance than one with a normal mean BMI,"[10] and that the marginal greenhouse gas emissions from food production and car travel totalled between 0.4 and 1 gigatonne of carbon per year per billion people, or 2.5% of global emissions. The authors tell the *Guardian* that "the heavier our bodies become the harder it is to move about in them and the more dependent we become on cars."[11]

A more recent study determined that the impact of obesity was slightly less at 1.6% of man-made emissions but provides more detailed data.[12]

> Compared with an individual with normal weight, researchers found an individual with obesity produces an extra 81 kg/y of carbon dioxide emissions from higher metabolism, an extra 593 kg/y of carbon dioxide emissions from greater food and drink consumption and an extra 476 kg/y of carbon dioxide emissions from car and air transportation. Overall, obesity is associated with approximately 20 percent greater greenhouse gas emissions when compared to people with normal weight.[13]

That's 1,150 kg total, equivalent to 46% of our annual carbon budget. Eating food we don't need from portions that are too big turns out to be one of the biggest sources of CO_2; talk about portion distortion!

In the end, it is clear that the amount that we eat can be as important as the amount that we waste. It's a demonstration again of sufficiency: how much do we really need? Perhaps it's time to downsize our dishware as well as our houses and cars and dig out Grandma's china. Bee Wilson wrote in the *Guardian*:

> If you want to see how inflated our portion sizes have become, don't go to the supermarket—head to an antique shop. You spot a tiny goblet clearly designed for a doll, only to be told it is a "wine glass." What look like side plates turn out to be dinner plates. The real side plates resemble saucers.[14]

Plastic and Food

For many years, I would show my students a box of Tropicana orange juice and a plastic jug of Ontario apple cider, explaining how one came from 100 kilometers away and sold at the farmers market, made from apples that were stored at slightly below room temperature, compared to orange juice, squeezed in Florida from Brazilian oranges, then stored in giant frozen blocks before it is packaged and shipped, refrigerated all the way from Florida to the refrigerated case in the grocery store.

But I still liked orange juice in the morning and didn't drink too much of it at a time, so would occasionally buy it in cardboard boxes. PepsiCo, owner of Tropicana, calculated the footprint of a 64-ounce box of juice to be 1.7 kg, with 60% being the juice, transport and distribution being 22%, packaging 15%, and customer use and disposal 3%; my 4-ounce shot is 106 grams of CO_2.[15]

But with plastics being so cheap, Tropicana switched to a clear plastic PET bottle that weighed 100 grams and, at 8 grams of CO_2 per gram of plastic, came out to 800 grams of CO_2 per bottle. I switched to the grocery store brand, but they went plastic within weeks to keep up with Tropicana. This is happening everywhere, as the price of PET continues to drop. So, it's back to local cider from the farmers market, in its plastic jugs. I asked the farmer if I could bring the old ones back for refilling, and he said "I don't want to go to jail. I can't do that!"

The Footprint of Fertilizer

Fritz Haber was a monster.[16] He invented explosives that are said to have lengthened the First World War by three years, and poison gas that killed thousands in that war and, after he was dead, millions of his fellow Jews in the Second World War. Yet in 1918, he won the Nobel Prize for Chemistry for his invention of the Haber-Bosch process of taking nitrogen out of the air, reacting it with hydrogen, and making it into ammonia, which is turned into nitrogen fertilizer, one of the legs of the green revolution. Half of the world's food is grown with nitrogen fertilizer.

$N_2 + 3H_2 => 2NH_3$

It takes a lot of hydrogen (3 molecules for every 1 of nitrogen), most of which is made from natural gas through steam methane reforming (SMR). Methane is the very definition of a hydrocarbon, with a formula of CH_4. The carbon is separated from the hydrogen and combined with oxygen to make carbon dioxide; the process is responsible for 1.8% of the world's CO_2 emissions.[17] In China, they gasify coal to make a third of the world's hydrogen,[18] with an even higher footprint. One source following the whole supply chain claims that "globally, ammonia production represents as much as 3 to 5 percent of carbon emissions."[19] Many people believe this isn't a problem; Vaclav Smil writes in *Energy and Civilization: A History* that it is a small price to pay.

> No other energy use offers such a payback as higher crops yields resulting from the use of synthetic nitrogen: by spending roughly 1% of global energy, it is now possible to supply about half of the nutrient used annually by the world's crops. Because about three quarters of all nitrogen in food proteins come from arable land, almost 40% of the current global food supply depends on the Haber-Bosch ammonia synthesis process. Stated in reverse, without Haber-Bosch synthesis the global population enjoying today's diets would have to be almost 40% smaller.[20]

But it is still a lot of energy, and a lot of CO_2. Fertilizer is made from ammonia, which is made from hydrogen, which is made from natural gas. That makes it a fossil fuel product; for every molecule of ammonia produced, a molecule of CO_2 is a co-product, so when we eat food made with nitrogen fertilizers, we are essentially eating fossil fuels. Not only that, but production is increasing worldwide, primarily driven by China and India as they increase their output and consumption.

What if we could do without?

The Organic Alternative

The alternative to eating fossil fuels is to eat organic food, where manure and other processes are used to add nitrogen to the soil. But can it do the job? Earl Butz, the Secretary of Agriculture for Presidents Nixon and Ford, and who had a way with words, noted on a TV show back in 1971:

> Without the modern input of chemicals, of pesticides, of antibiotics, of herbicides, we simply couldn't do the job. Before we go back to an organic agriculture in this country, somebody must decide which 50 million Americans we are going to let starve or go hungry.[21]

Organic farmers have been quoting him ever since, as they try to prove him wrong. A 2017 study lead by Adrian Muller developed strategies for feeding the world with organic agriculture, claiming that "a worldwide conversion to organic farming can contribute to a comprehensive and sustainable food system, if combined with further measures."[22] They acknowledged that organic produces lower yields and would require more land, which could lead to increased deforestation.

To feed the world with an organic diet would require a number of "further measures" or concurrent changes, including increasing the supply of organic food by taking over land used for raising animals, reducing demand for animals by promoting a healthier diet, and significantly reducing food waste.

Mark Lynas of the Cornell Alliance for Science thought this to be questionable at best.

> How do the authors achieve their headline conclusion? By combining a worldwide conversion to organic agriculture with a heroic parallel worldwide conversion to vegetarianism, allowing them to assume (in some scenarios) a 100 percent

> reduction in land-area competition from animal production. This is combined with a similarly heroic 50 percent reduction in global food waste.[23]

He considers this to be an impossible goal: "However desirable vegetarianism might be both environmentally and for human health, global consumption of animal products is going up not down as developing countries achieve higher standards of living."[24]

The problem is, we really don't have a choice.

The Whole Food System Needs an Overhaul

A study released in November 2020 examined the full scope of the carbon emissions from the worldwide food system and concluded that its carbon footprint is so big that "even if fossil fuel emissions were immediately halted, current trends in global food systems would prevent the achievement of the 1.5°C target and, by the end of the century, threaten the achievement of the 2°C target."[25]

The global food system generates emissions every step of the way: land clearing and deforestation releases CO_2 and nitrous oxide (N_2O); production and use of fertilizers releases CO_2 and methane; cows, sheep, and pigs release methane during digestion; rice paddies and livestock manure both release methane and nitrous oxides; and then there is all the equipment, the trucking, the supply chains. The researchers didn't even include transportation, the cold chain, processing, packaging, and retail and still came up with emissions of 16 billion tonnes of CO_2 per year. They note also that emissions will continue to grow as countries become more affluent and if we keep doing business as usual.

Instead of business as usual, the researchers call for five measures that could reduce emissions by up to 48%:

- Adopting a plant-rich diet such as a Mediterranean diet or the EAT-Lancet diet (also called the Planetary Health Diet), "moderate amounts of dairy, eggs and meat";
- Reducing the amount that we eat: "adjusting global per capita caloric consumption to healthy levels";
- Improving yields through crop genetics and better farming practices;

- Reducing food loss and waste by 50%;
- Reducing the use of nitrogen fertilizers through more precise use.[26]

I discussed earlier the carbon footprint of meat and dairy and was surprised that the researchers didn't recommend cutting them out completely. I contacted the study's lead author, Michael Clark, who responded:

> You're correct that we did not include a vegetarian or vegan diet, but I also wouldn't say that the EAT-Lancet diet is far more moderate than these. The EL diet allows for ~14 g red meat/day, with slightly more poultry and fish. Compared to current diets in many countries, meeting the EL diet would still require a very very large change from current dietary choices. From a psychological perspective, communicating "eat less meat" seems to be a more effective way to get people to shift their dietary habits than is "eat no meat."[27]

Clark and his team note that other benefits accrue if these kinds of changes were made, including reductions in pollution from fertilizer runoff, improved biodiversity, decreased land-use change, and "if dietary composition and caloric consumption are improved, reduced prevalence of obesity, diabetes, heart disease, and premature mortality." It all sounds like a win-win situation to me. The study concludes: "Time is of the essence in addressing GHG emissions. Any delays will necessitate more ambitious and expeditious implementation of emissions reduction strategies if global temperature targets are to be met."[28]

The Cold Chain and Local Food

The authors of the global food system emissions study conspicuously excluded "emissions from transportation, processing, packaging, retail, and preparation, which in total account for a minor fraction (~17%) of total food system emissions." To me, 17% certainly doesn't seem minor. In the footnotes, the authors reference as their source an article by Joseph Poore and Thomas Nemecek[29] whose work is really the foundation of the understanding of the carbon footprint of food and what I have been relying on throughout this exercise. Hannah

Ritchie of Our World in Data uses their research to conclude that transportation is a minor component as well, writing:

> Eating locally would only have a significant impact if transport was responsible for a large share of food's final carbon footprint. For most foods, this is not the case. GHG emissions from transportation make up a very small amount of the emissions from food and what you eat is far more important than where your food traveled from.[30]

There is no argument here; what you eat is more important than where it came from. Richie also references the work of Christopher Weber and Scott Matthews and their article "Food-Miles and the Relative Climate Impacts of Food Choices in the United States," noting that "their analysis showed that substituting less than one day per weeks' worth of calories from beef and dairy products to chicken, fish, eggs, or a plant-based alternative reduces GHG emissions more than buying *all* your food from local sources."[31]

This did not sit well with me. My late dad was in transportation, and I grew up around containers and "reefers" or refrigerated trucks, and this didn't pass the smell test. And sure enough, when I dug into the data, all Weber and Matthews measured were the food-miles, the distance the food travelled, which they converted to CO_2 emissions using accepted data for different forms of transport. They do note that "there are also deviations from the average energy intensities per km used here; for example, refrigerated trucking and ocean shipping of fresh foods are more energy-intensive than the average intensity of trucking or ocean shipping."[32] However, they never really get into how much that is, suggesting that it doesn't change their argument that food-miles don't matter much, and the words "cold chain" never come up. They never did that math.

I assigned this problem to one of my students at Ryerson University, Yu Xin Shi, who found research confirming that there is much more to the cold chain than just the question of food-miles. One study reports that "overall 15% of the world's energy production is used to power cold chains and cooling systems which still depend on fossil fuels."[33] Jean-Paul Rodrigue and Theo Notteboom, authors of *The Geography of Transport Systems*,[34] define it as a "process since a series of tasks must be performed to prepare, store, transport, and monitor temperature-sensitive products."[35] It includes:

- *Cooling systems* to get the food down to the appropriate temperatures for processing, storage, and transportation;
- *Cold storage*, "Providing facilities for the storage of goods over a period of time, either waiting to be shipped to a distant market, at an intermediary location for processing and distribution, and close to the market for distribution."
- *Cold transport*, moving and keeping food at the right temperature and humidity;
- *Cold processing and distribution*, consolidating and deconsolidating loads.[36]

The cold chain is a big operation, and it is getting bigger all the time. The authors write:

> Increasing income levels are associated with a change in diet with, among others, growing demand for fresh fruit and higher value foodstuffs such as meat and fish. People with higher socioeconomic status are more likely to consume vegetables and fruit, particularly fresh, not only in higher quantities but also in greater variety. Consumers with increasing purchase power have become preoccupied with healthy eating. Therefore, producers and retailers have responded with an array of exotic fresh fruits originating from around the world.[37]

So those food-miles are in fact just a part of the bigger picture. And they might not even be accurate; according to a study on food transport refrigeration, "greenhouse gas emissions from conventional diesel engine driven vapour compression refrigeration systems commonly employed in food transport refrigeration can be as high as 40% of the greenhouse gas emissions from the vehicle's engine."[38] The emissions per food-mile could be almost double.

According to the United Nations Environment Programme document *Sustainable Cold Chain and Food Loss Reduction*, transportation directly contributes 1.2 gigatonnes of CO_2e annually from burning diesel, and is responsible for up to 7% of the world's hydrofluorocarbon emissions because of "high refrigerant leakage and poor end-of-life disposal, contributing as much as 4% of the total GHG emissions of transporting all freight (refrigerated or not)."[39] HFC refrigerants have up to 22,800 times the global warming potential of carbon dioxide; surely, we want to minimize this by using less refrigerated transport.

This is why local food is back on the menu: food-miles are a poor reflection of the true impact of the cold chain. Notwithstanding the numbers from these various studies and the Our World in Data gang, I am sticking with the advice from Italian communalist Lanza del Vasto: "Find the shortest, simplest way between the earth, the hands, and the mouth." Avoid the cold chain wherever possible, and if you can't, keep it as short as possible.

What Do We Learn from the Spreadsheet?

It becomes very clear that I shouldn't have eaten that beef with broccoli in that wonderful Chinese dinner that we ordered in. In fact, we shouldn't have ordered in at all, with all the plastic and foam containers and, of course, the delivery guy's car. But my wife had spent the day canning tomatoes and was exhausted, and the kitchen was an oven. But it became clear when tracking my diet that red meat is off the menu and dairy is following closely; even my beloved morning cappuccino is 235 grams of CO_2 because of the milk.

What Can We Do?

We have supersized everything we own and supersized ourselves, and it has a surprisingly large carbon footprint. In the end, this entire chapter can be summarized with Michael Pollan's famous seven words: "Eat food, not too much, mostly plants." But more explicitly, for a low-carbon diet:

- If you can, go vegan and eliminate all meat, fish, and dairy.
- If (like me) going vegan is too hard, red meat, lamb, and shrimp are off the menu.
- Eat less dairy, fish, chicken, and pork; they have a fraction of the footprint of beef, but it's still three times as big as the footprint of plants.

Table 5.1. Calculations for food and alcohol

Item	Source	Unit	CO_2/ unit	E stimates/ day	Nov 9	Nov 25	Nov 26	Nov 27	Nov 28	Nov 29	Nov 30
Food											
veg breakfast	est JC mortreaux			350.00	800	595	410	1,050	1,050	525	525
veg lunch	est JC mortreaux			500.00		500	500		800	600	525
veg dinner	est JC mortreaux			500.00			500	500		500	
beef	Omni	serving		7,200.00					1,800		
pork	Omni	serving		950.00				475		950	950
chicken	Omni	600		800.00		4,600					
fish	Omni	serving		1,100.00							
cheese	Poore	per 100 g		1,100.00							
milk	Poore	per liter		3,200.00		800					
snack				300.00							
Alcohol											
red wine	food climate research	glass		275.00					X		550
beer	food climate research	pint		330.00							
martini	food climate research	double		123.00							123

- Shorten the food chain and the cold chain; buy local and seasonal and cook it yourself instead of buying processed or prepared foods.
- Learn to love leftovers and throw out as little as possible.
- Manage portion sizes and don't cook more than you need. It is surprising how big the impact of this is.
- If you do takeout or delivery, walk or bike to pick it up yourself or find out who delivers by bike.

– 6 –

How We Live

From the 1.5-degree lifestyle report: "housing infrastructure and supply of utilities, e.g., construction, maintenance, energy use and water use."

Where We Live and Work

I noted earlier that "land use, transportation, *and energy* are the same thing described in different languages," describing how the availability of a dependable electricity supply led to the development of the electric streetcar, which led to the urban form of the streetcar suburb and the design of the house where I live.

But when you start looking at the footprint of our homes and where we live, you cannot separate it from where we work and how we get between them; it's all one picture, the same thing described in different languages. That's because we are all living in a world that was constructed in a different era; it was built in the Second Industrial Revolution that ran from the 1870s through the early 1900s. As Smil notes in *Growth*:

> Gordon (2016) concludes that the second industrial revolution of 1870–1900 (with its introduction of electricity, internal combustion engines, running water, indoor toilets, communications, entertainment, launching of oil extraction and chemical industries) was far more consequential than both the first revolution (1750–1830, introducing steam and railroads) and the third (begun in 1960 and still unfolding, with computers, the Web and mobile phones as its icons).[1]

The Third Revolution has recently been unfolding with a bang, thanks to the COVID-19 pandemic, but the way most of us live, work, and get around is a construct from the Second. Prior to the 1870s, North America was overwhelmingly agrarian, with the population spread far more evenly than it is now, with flourishing small towns popping up everywhere with independent shops serving the local farming communities. When the railways came, they fed into bigger administrative centers, where many were employed as scriveners (copiers) and book-keepers. Offices existed but they were generally very small.

But two forms of energy changed everything: steam power for the railways and electricity for the telegraph. According to Margery Davies, writing in *Woman's Place Is at the Typewriter: Office Work and Office Workers, 1870–1930*, this led to "massive consolidation of businesses... [with] giant corporations integrated vertically and horizontally in the merger movement that swept through industry during the 1890s. In the steel, oil, tobacco, food, and meat-packing sectors, to name just a few, such corporations enjoyed virtual monopolies."[2] Transactions could no longer be handled face-to-face, and accurate records became more important. Davies notes that your small-town butcher might just need simple records, but a big meat-packer needs information from all over the country. Scriveners couldn't keep up.

As so often happens, technology pops up to fill the need. The typewriter had been around for decades, but nobody would invest in its production because nobody saw the need for it. As Davies writes, "the potential value of a writing machine was not readily apparent to businessmen who ran small offices with a few clerks and a relatively small amount of paperwork."[3] In 1873, the gun-maker Remington bought the rights to a new typewriter design from the so-called Father of the typewriter, Christopher Sholes, and by 1885 it became a staple of the office. But Davies makes the point that is relevant to almost all inventions, which is that they do not just pop up of nowhere:

> It is clear that in the development of the typewriter, changes in the organization of capitalism gave rise to technological innovation, rather than the reverse. Inventors had been experimenting with writing machines for over 150 years before the Remington company started mass production of the Sholes

> typewriter. It was only in the 1870s, with the first indications of the expansion of offices and the growth of office work, that any capitalist firm was willing to invest in the manufacture of writing machines. It was not until the 1880s, when offices grew by leaps and bounds, that the typewriter began to sell. Rather than causing change, the typewriter followed in the wake of basic alterations in capitalism.[4]

As the offices flourished, they needed stenographers who could take dictation and they needed typists. The demand was so great that there were not enough men to do the job (and many didn't want to be stuck in the same job with little chance of advancement), so companies started accepting women; there were more female and literate high school graduates who were willing to learn how to type, and they got paid less, too. With the industrialization of farming, people flocked to the cities where these jobs were, where women could significantly contribute to the family income. With typing and carbon paper, there was an explosion in paper consumption and the invention of the vertical filing cabinet, and the need for ever more office space to keep it all handy, central, and accessible. But it all had to be close to where the workers lived, so the elevator was put to work (it had been around for a while too) so that buildings could go up and stack more people more closely together. And in the space of just a few decades, between 1870 and 1910, we pretty much got the cities we have today, with office buildings and apartments and suburbs, subways and streetcars, all running on coal and steam and electricity and telephone wires.

This is an example of why you cannot separate how we live from how we work and how we get between the two, and how you can't deal with one aspect of our lives without dealing with the others. Look what happened when gasoline became the transportation energy of choice, eclipsing steam and electricity.

The Problem with Cars

Cars started pushing horses, pedestrians, pushcarts, and everyone else off the roads after the First World War, but most people still came downtown to work and the basic urban patterns didn't radically change until after the Second World War and the start of the nuclear age. Defence planners quickly realized that a single bomb dropped on New York City could take out the entire American economy, with so many businesses concentrated in one place. Shawn Lawrence Otto describes in *Fool Me Twice* how the Commander in Chief of the Continental Air Defense Command told a conference of mayors in 1954:

> Your city means everything to you, everything to the people who live in it, and everything to me. To our possible enemies, however, who sit down at their planning tables to compute a schedule of take-off times for their existing bombing fleets, the hundred biggest cities represented here by you do not mean historic streets and beautiful parks, school systems in which you have pride, or the churches which are your fountains of faith. They may mean to them only those aerial forces and weapons required to produce the 100 pinpointed minutes of atomic hell on earth necessary for their destruction.[5]

Decentralization was the answer, and the car was the key because it could go anywhere. The American government built the National Interstate and Defense Highway System to move people quickly; it was the single biggest investment ever made by the U.S. Government.[6]

National Industrial Dispersion Policies encouraged companies to move out of big cities and build office parks in the country, and brought in mortgage guarantee programs so that white middle-class people could buy low-density housing in vast new suburbs and still have enough money to buy the cars they needed to get between everything, on subsidized roads with subsidized gasoline.[7]

It was a beguiling lifestyle, and car-dominated suburbia was copied everywhere and kept going long after the nuclear threat receded. The problem is that it all depended on lots of cheap fossil fuels, to heat and cool the houses that kept getting larger, along with the cars that grew into SUVs, driving further to the schools and the big-box stores in the outskirts. Now close to 75% of the North American population lives this way, and it is very hard to fix.

Learning from Saskatchewan

When the oil crisis hit in 1976, everyone suddenly got serious about energy conservation and solar-powered houses became all the rage.

"Mass and glass" was a popular approach, with sun streaming through windows and warming floors, special heat-absorbing Trombe walls and even drums of water. The idea was that the sun would supply the heat and the mass would store it.

Meanwhile, in chilly Saskatchewan, the government wanted on the solar bandwagon and asked the Saskatchewan Research Council (SRC) to design a solar house appropriate for the province. But there is not much sun in Saskatchewan during winter, and engineer Harold Orr and his team calculated that the amount of storage that would be needed was vast (enough water to fill two Olympic-sized swimming pools). So, Orr and his team flipped the problem from worrying about getting enough *supply* to, instead, reducing *demand*. This was a critical insight that is still controversial today.

Orr's team increased the insulation way beyond what anyone had considered before, six times what is standard today. They designed a simple boxy form, shrunk the windows, and installed shutters to keep the heat in at night (triple-glazing didn't exist then). They sealed the building to almost eliminate leakage and invented a heat exchanger to bring in controlled amounts of fresh air and pre-warm it with the exhausted air.

But insulation isn't sexy. You can't see it. Even the Saskatchewan government was unimpressed and insisted that they add solar collectors for hot water, a system that cost almost as much as the house. Harold Orr told the CBC: "It only cost $45 a year to heat [the house], but the maintenance costs on the [solar] panels were nearly $10,000."[8] In 1978, Orr's associate Robert Dumont wrote a paper comparing the Saskatchewan Conservation House to a "mass and glass" house, and it was immediately clear which was more efficient, more comfortable, more affordable, and easier to maintain.

It was then all promptly forgotten about, until Wolfgang Feist studied it and other super-insulated houses and came up with the Passivhaus or Passive House concept, which has all the attributes of the house first built in Regina, Saskatchewan.

Many insights can be gleaned from the Saskatchewan Conservation House and the Passive House concepts, but the key one is to always look at the *demand* side first. This applies to everything: Reduce demand for energy with insulation. Reduce demand for materials by building less. Reduce demand for transportation by building closer together. If you reduce demand first, coming up with the supply is so much easier and cheaper.

So, if you are going to build a society based on reduced demand, you might start by demanding that every new home be built to extremely high standards of energy efficiency at streetcar-suburb densities in walkable neighbourhoods. Or one could go full Vienna and just say no to any single-family housing, everyone can live in lovely apartments built after the subway line and bike lanes are finished. Missing Middle Housing is all the rage right now, with zoning bylaws being changed to allow duplexes and triplexes. I always pitched what I called "the Goldilocks Density" and described it in the *Guardian*:

> Dense enough to support vibrant main streets with retail and services for local needs, but not too high that people can't take the stairs in a pinch. Dense enough to support bike and transit infrastructure, but not so dense to need subways and huge underground parking garages. Dense enough to build a sense of community, but not so dense as to have everyone slip into anonymity.[9]

Studies have shown that density is the single biggest determinant of carbon footprint. A few years ago, the Urban Archetypes Project[10] for Natural Resources Canada looked at 31 neighbourhoods in 12 Canadian cities and found that people who lived in apartments over stores in Calgary's Mission district or in walk-ups in Ottawa had footprints that were a fraction of the size of those who lived in the suburbs because, even though their homes were between 50 and 100 years old and had no insulation or crappy windows, they basically lived in much smaller spaces and didn't drive everywhere.

Dense also doesn't necessarily mean tall; go to Montreal and visit the Plateau district with some of the densest housing in North America, all in three-storey buildings. They are probably the most efficient house forms I have seen anywhere; every apartment has front and rear windows and no space lost to circulation and hallways because there are crazy steep and twisty death-trap stairs out front.

It turns out that at the turn of the twentieth century the Montreal planners and politicians were concerned about everything being too dense and tight and demanded a ten-foot setback from the property line so that people could have a little lawn or garden. Developers being developers, they immediately thought that they were not going to waste all that space, so they put the stairs to the upper apartments there. But ten feet isn't quite enough to do a safe straight stair, so they twisted and contorted and steepened them to squeeze them in. They are glorious and ridiculous, especially given Montreal's winters; had the planners demanded twelve feet, it would have been fine. But they still demonstrate that you can design comfortable, desirable housing in walkable communities where everyone has their own front door. There are versions being designed today that have safer stairs and are fully accessible via outdoor corridors.

Density can be achieved in many ways, without you even seeing it. A few years ago, after our two children had moved out, I wanted to sell our big leaky house and move to a condo, but my wife, Kelly, wanted to stay put. Finally, we agreed to duplex the house, turning it into two units. We fixed the leakiest part, the rear, but didn't insulate or change the windows in the main part of the house. We caulked and sealed as best we could. But we cut our own living space to less than half of what it was and rented out the upstairs to one of our daughters at market rates, which covered the cost of the renovation. The heating and electrical costs are lower than they were, thanks to the sealing and the fixing of the rear, but there are now five people in it instead of two. The per capita emissions have dropped to less than half because that is what we are working on reducing, our per capita emissions. I am now sharing my electric and gas bills, and much of my carbon footprint from housing, with another family.

The Fuel Problem

We have a gas stove in the kitchen and a gas boiler delivering hot water to our 100-year-old radiators and to our domestic hot water tank. Many years ago, I thought this made total sense. If I want a bowl of pasta, I am boiling water directly from gas, whereas if I used electricity, I was really just letting someone else boil water with coal, gas, or nukes to make steam to spin a generator to push electricity down a wire—losing bits of it along the way to generator friction, inefficient insulators, resistance in the wire, transformer heating—finally getting it to my house where I push it through a coil to.... boil water.

But much has changed over the years. Where I live in Ontario, Canada, we are blessed with Niagara Falls and other hydroelectric sources developed more than 100 years ago. Fifty years ago, the Province started investing in a fleet of nuclear reactors that cost way too much to build but now deliver lots of clean power. Ten years ago, Premier Kathleen Wynne got rid of the last coal-fired power plants and controversially replaced them with peaker gas-fired plants that would turn on only when they were needed; now fossil fuels power only 4% of our electricity supply.

All over the world, electricity grids are getting greener as more renewables come online and more coal plants are shut down. Battery technology is getting so much better and cheaper that it is replacing the gas peaker plants. The electrical grid is getting better every day.

The gas grid is not. The utilities promise more biogas and are now touting hydrogen, but they are both fantasies. That's why in the end we have to electrify everything and get our homes off gas. But this is also hard because, where I live, gas is really cheap, thanks to fracking in the US, and electricity is really expensive, thanks to "stranded debt" from all those overpriced nukes. Had I replaced my gas boiler with an electric one, my operating costs would triple. Had I invested in a ground source or CO_2 air source heat pump system, my capital costs would triple or more. I would have had to do a deep retrofit, insulating the entire house, to get the heating load down to where I could run it with an air source or single CO_2 heat pump, in which case my renovation costs would have doubled and I would have lost all that old-house charm. So now I am locked in to gas. This is a problem with almost every existing house in Canada and the northern United States where heating is the main use of energy.

How Do We Fix What We've Got?

The question of what to do with the millions of existing gas-burning and carbon-spewing homes is one of the most vexing challenges we face. Renovations of single-family houses are expensive and disruptive; building trades experienced in energy retrofits are few and far between. There is not even a consensus about the building science, about what to do.

There is also a tendency to throw hardware at the supply side of the problem, from solar panels to more efficient heating and cooling systems or smart thermostats, rather than dealing with the demand side, as Harold Orr taught with the Saskatchewan Conservation House.

During the energy crisis in the seventies, the standard approach was to add insulation wherever you could, as much as you could. This might involve gutting the house to add insulation to the exterior walls, wrapping the entire house in foam and stucco, and changing all the windows. The replacement window salespeople were particularly aggressive, promising that they could cut your heating bills in half. In the end, many people were disappointed, having spent small fortunes and reaping small savings, far less than promised. This is still standard practice with unsophisticated builders and clients, and much of the money and effort was misdirected and wasted.

Then the blower door came from Sweden to North America in 1979, and it changed the way we looked at buildings. Actually, it was a blower window, but researchers quickly realized that door sizes are standardized, so a blower door would be more flexible. The fan in the blower door could pressurize or depressurize a building while being monitored to see how much air was moving through at a standard pressure differential between inside and outside, usually 50 pascals.

Researchers were blown away, so to speak, by their testing results: heat loss through leakage was far more significant than anyone suspected. Building science changed, if not overnight, but rapidly as they realized that airtightness was as important as insulation.

In the early 80s, Harold Orr and Robert Dumont of Saskatchewan Conservation House fame did what has become known as a "chainsaw retrofit" of an existing sixties bungalow. They chainsawed off the roof overhangs and wrapped the entire house in insulation. However, when Orr studied the results, where the money went and what made a difference, he concluded that wrapping and residing a house doesn't make much sense. He was interviewed by Mike Henry in 2013:

> "When you put Styrofoam on the outside of a house, you're not making the house any tighter, all you're doing is reducing the heat loss through the walls. If you take a look at a pie chart in terms of where the heat goes in a house, you'll find that roughly 10% of your heat loss goes through the outside walls." About 30 to 40% of your total heat loss is due to air leakage, another 10% for the ceiling, 10% for the windows and doors, and about 30% for the basement. "You have to tackle the big hunks, and the big hunks are air leakage and uninsulated basement."[11]

If every house in the country had a blower door test and a few shots with a thermographic camera in winter and just a few tubes of caulk to seal up the leaks, it would lead to significant reductions in carbon emissions: by Harold Orr's calculations, perhaps 30%.

But building science is complicated; if you have a tightly sealed house with poorly insulated walls, you can get condensation and mould in cold weather. You can run into indoor air quality problems with CO_2 from people, and NO_2 from gas appliances. Bathroom and kitchen fans or clothes dryers can create negative pressure and cause backdrafts from furnaces, bringing in carbon monoxide. The leaky house may have been full of cold dry air, but it was the unscientific way of getting fresh air; that has to somehow be replaced with an engineered source of fresh air. Often this is just a vent to the outdoors; more sophisticated but expensive heat recovery ventilators are becoming more common in renovations.

Corey Diamond of Efficiency Canada suggests that I am oversimplifying and that it isn't so simple as just caulking everything, telling me in an interview in December 2020: "Every house is different, you have to treat it as a system, not just look at the furnace or the windows. Where is it leaking? What needs to be fixed? What gives the biggest bang for the buck?"

The other extreme is Energiesprong, a Dutch concept where the entire house is wrapped in a prefabricated wall made to order, complete with new windows and doors and a thick wrap of insulation, sort of a modern high-tech version of the chainsaw retrofit. Furnaces are replaced with little heat pumps, and the entire process takes just a couple of days. The Dutch are planning to do every house in the country, all two million of them, by 2050, at a cost of billions.

Many groups are trying to bring Energiesprong to North America, but the houses are bigger than they are in the Netherlands, and are usually detached and with more complicated geometries, so the costs go up dramatically; we could be talking $100,000 per house. Just in Canada, almost eight million houses[12] need fixing; multiply that out and you have a serious number of zeros.

It's all made more complicated right now because gas is so cheap; the payback period on anything other than caulk and sealing is almost forever. There is no economic incentive for a homeowner to renovate just for energy savings, and it is hard to imagine our governments putting on a big enough carbon tax to change that.

The British seem to be throwing up their hands and are trying to decarbonize the fuel; the Committee on Climate Change[13] is proposing that hydrogen be mixed in with natural gas, eventually replacing it. This is as much of a fantasy as Canada investing eight trillion dollars in fixing its housing stock.

Caulk won't ultimately get us where we need to go by 2050, but it is a good starting point. When a blower door test was done on my house before we renovated a few years ago, the house was so leaky that the fan in the door couldn't even get to a 50-pascal pressure differential. Through careful sealing and caulking and the addition of window inserts on the 110-year-old double-hung windows, we were able to reduce our gas consumption for heating by 30%. And it all starts with that big red blower door.

There is not enough money in world to do an Energiesprong renovation to every house in North America, but there is enough to deal with what Harold Orr called the "big hunks," the air leakage, basements and attics if accessible, reducing energy requirements for heating and cooling by up to 70%. Then you can make up the difference with a modern air source heat pump. The house may not be net zero energy, but in many parts of North America, the house will be zero carbon and the grid is getting cleaner every day. This is the only way we can possibly deal with the millions of houses that have to be fixed.

Awash in Water and Waste

While tracking my carbon footprint, I have been monitoring the number of baths and showers because of the incremental footprint of heating the water. It never occurred to me to actually monitor the amount of water used; I live by Lake Ontario and never thought much about my water usage, just the energy for heating it. But in fact, the embodied energy of treating and pumping water is significant, according to a 2011 study, as much as a third of a municipality's energy usage.[14] The average American family uses 100 gallons (379 litres) of drinking water per day, with an average embodied energy of 1.1 kWh,[15] which in the US has a footprint of about one pound of CO_2. That doesn't sound like much, only 1.2 grams per liter, but it adds up fast, and it seems particularly wasteful just to use so much of it to flush toilets.

After all that expensively cleaned and pumped water is used, the moderately dirty graywater from sinks and showers is mixed with the blackwater from toilets, contaminating it all, and then run through networks of increasingly giant pipes to sewage treatment plants, where the processing of all the poop releases nitrogen dioxide and methane. According to the EPA, the energy consumed treating water, moving water, and then treating the wastewater could amount to around 3% to 5% of total global energy consumption.[16] Paula Melton of BuildingGreen asks: "Given the amount of energy it takes to move potable water around, does it make better energy sense to collect and treat rainwater, graywater, or blackwater on individual building sites?[17]

The answer depends on scale; this is far easier to do on a single-family house with a site big enough to process wastewater. The Living Building Challenge (LBC) certification system demands it, but one can argue that drinking water is something that is best managed collectively.

You can visit the Bullitt Center in Seattle, one of the world's greenest buildings, and there is a fancy and expensive water system in the basement, separating the water from the dirt and bird poop collected from its solar panel roof. It's then chlorinated, since Seattle demands it, and then dechlorinated, because the LBC bans chlorine.

Meanwhile, everyone else in Seattle is drinking what is considered the "cleanest and best-tasting drinking water in the nation," collected from snow-fed mountain rivers and constantly monitored by water quality experts.

Wastewater is another story, a tragedy of errors, incompetence, and misguided thinking that got us into the mess we are in today. Nobody actually planned any of this; it all happened by accident, and it is a story that starts in 1854, and it costs us mightily in our carbon footprints. That's when Dr. John Snow made the connection between a cholera epidemic in Soho (then not such a fancy part of London) and a pump where the locals got their water. People in cities often used vault privies, or underground stone tanks that would fill with poop and get emptied occasionally by night-soil men who carted it off to the fields to be used as fertilizer. A leaky vault contaminated the groundwater supplying the pump. Snow didn't know about bacteria but did make the connection that poop + drinking water = death.

Soon water companies were laying pipe all over London to supply supposedly safe water to peoples' homes, usually sticking a faucet in the kitchen or a washstand in the bedroom. When people had to carry the water home in pails, they didn't use very much of it, but now they could use as much as they wanted, and they used a lot. Abby Rockefeller wrote in "Civilization & Sludge: Notes on the History of the Management of Human Excreta":

> The system of cesspools and vault privies, which had been to some extent effective in avoiding pollution of waterways through their periodic cleanout by scavengers and the at least partial returning of human manure to farms, was overwhelmed by the pressure created by the new availability of running water.[18]

Toilets were invented earlier but were useless without water; now they were being installed in closets (WCs) even though there were no sewers. Everything was just dumped in drainage ditches that meandered down to the Thames, which got so filthy that Parliament couldn't meet during The Great Stink of 1858. That instigated a massive sewer building project that took all the waste downstream of

London and dumped it back into the Thames there. Engineers in Europe and the US watched all this and wondered if it wasn't a mistake. Abby Rockefeller wrote in "Civilization & Sludge":

> The engineers were divided between those who believed in the value of human excreta to agriculture and those who did not. The believers argued in favor of "sewage farming," the practice of irrigating neighboring farms with municipal sewage. The second group, arguing that "running water purifies itself" (the more current slogan among sanitary engineers: "the solution to pollution is dilution"), argued for piping sewage into lakes, rivers, and oceans. In the United States, the engineers who argued for direct disposal into water had, by the turn of the 19th century, won this debate. By 1909, untold miles of rivers had been turned functionally into open sewers, and 25,000 miles of sewer pipes had been laid to take the sewage to those rivers.[19]

In Asia around this time, people in cities would poop into terra cotta pots, which they would put out on the street in the morning. In Shanghai, there was a network of canals that would carry what looked like gondolas of poop upstream to fertilize the farmland. It was in such demand that in Japan people got paid for their poop; of course, the rich got more money because their poop was richer too because of the higher-quality food that they ate. All of this disappeared in the twentieth century; Fritz Haber delivered nitrogen fertilizers without the mess or smell.

So now we have a system where we generate 3% to 5% of our carbon emissions cleaning water that we use to flush away useful potassium in pee and nitrogen in poop, while we generate another 3% to 5% of our emissions making fertilizer full of nitrogen and potassium.

It doesn't have to be this way; we could get off-pipe with composting toilets and save all those nutrients, clean up our rivers, and save billions of dollars on waste management, not just in homes. The C. K. Choi Building at the University of British Columbia[20] has Clivus Multrum composting toilets servicing the entire 30,000-square-foot building. The six-storey Bullitt Center has Phoenix composters in the basement, right next to its silly water supply system. The washrooms in the Bullitt Center are the sweetest smelling ones I have ever used. Composting toilet systems are all under negative pressure, sucking air down through the toilet, so they never smell like bathrooms.

We could all have this, and unlike my composting toilet in my cabin, city people would have a service that would pick it up once or twice a year. And given the North American diet, it would probably be high-quality poop.

What We Build With

There is also the question of the materials we build with, where outside of the single-family housing world, we use vast amounts of steel and concrete. The making of steel and cement each contribute about 6% of the annual production of CO_2. The issue with both materials is that the CO_2 isn't just generated from heating it or melting it, but about half their emissions are directly related to the chemistry of the process.

Concrete

The key component of cement is lime (calcium oxide), which you get by applying heat to calcium carbonate, basically limestone.

$$CaCO_3 + \text{heat} > CaO + CO_2$$

You can't do anything about the chemistry. You can use less lime and substitute fly ash and pozzalan (what the Romans used, basically volcanic ash) and reduce the carbon footprint somewhat. But it is the fundamental nature of the material that making it emits CO_2.

Steel

Steel also has a chemistry problem: it's made from iron, which is made from hematite or iron oxide, which is found in nature. This is melted in a blast furnace, where pure carbon, in the form of coke, is thrown in; the carbon combines with the oxygen and is emitted as CO_2, the result being steel.

$$Fe_2O_3 + 3\ CO \text{ becomes } 2\ Fe + 3\ CO_2$$

There is progress in making better steel with a lower carbon footprint. German steel maker ThyssenKrupp is injecting hydrogen instead of coal, which bonds with the oxygen to make water instead of CO_2. Hydrogen is currently made from natural gas (in a process that emits CO_2) but could eventually be made using electricity, leading to "green" steel if you ignore all the mining and all the melting of the steel.

Aluminum

It's the same with aluminum. Historically, the electricity needed to break the bond in alumina, or aluminum oxide, between the aluminum and oxygen was the big carbon emitter because it was generated with coal. However, prior to the Second World War, the Columbia River was dammed, and the Tennessee Valley Authority built many dams, all primarily to provide power for aluminum production for airplane manufacture. Canada, Norway, and Iceland all built massive hydroelectric generating facilities to produce aluminum that's called "green" because of the clean electricity. But there is still CO_2 in the chemistry; the anode supplying the electricity is made of pure carbon, which combines with the oxygen to make carbon monoxide (CO) and CO_2.

$$AlO_3 + 3/2C \text{ plus 3 electrons becomes } 2Al + 3/2CO_2$$

There's new improved technology that replaces the carbon anode with a proprietary material that releases oxygen, not CO_2, and which has been given a big boost by Tim Cook of Apple. However, you still can't call aluminum green because mining bauxite is hugely environmentally destructive, and separating alumina from bauxite ore is still a messy process. And while aluminum is easy to recycle, demand is always increasing so there is still need for virgin aluminum. You also cannot build a computer or an airplane out of post-consumer recycled aluminum; there are critical specifications for the alloys. (Apple's much-promoted use of recycled aluminum only uses pre-consumer waste from their own manufacture, which is kind of like eating leftovers; it would have been better had there been no waste in the first place.) Instead, we have to consume less of the stuff; Carl Zimring wrote in *Aluminum Upcycled* (his preferred word for recycling of aluminum) that we have to reduce our use to the point that we don't need to make the new stuff:

> As designers create attractive goods from aluminum, bauxite mines across the planet intensify their extraction of ore at lasting cost to the people, plants, animals, air, land and water of the local areas. Upcycling, absent a cap on primary material extraction, does not close industrial loops so much as it fuels environmental exploitation.[21]

Steel, aluminum, and concrete, three of the fundamental materials used in construction, are all made through chemical processes that include the production of CO_2 and releasing it to the atmosphere.

Plastic

The construction industry is one of the largest consumers of plastic, which I have described as a "solid fossil fuel." Most green building certification systems are trying to get plastics (particularly polyvinyl chloride [PVC], also known as vinyl) out of buildings because of toxicity, but the industry is fighting back, and making safer vinyls. However, it doesn't change the fact that it is made from chlorine and ethylene (fractionally distilled from oil) naptha (from natural gas) or in China, where about half of the world's PVC is made, from coal, in a process described as "very energy intensive with much waste generated."[22]

Wood

Wood is the complete opposite. Carbon dioxide is absorbed from the atmosphere and combined with hydrogen derived from water to make complex molecules that work like natural structures, with lignin acting as the concrete, resisting compression, and the cellulose fibres handling the tension. That carbon that is drawn out of the air is stored in the wood until it rots or burns.

Wood has been used for construction since people started building, and most of North America's single-family housing has been built out of it. However, in the last 25 years, there has been a revolution in wood construction, as mass timber has been developed to replace concrete and steel. It comes in wonderful new high-tech forms from cross-laminated to dowel-laminated to laminated-veneer, all taking little bits of wood from small trees and holding them together with glue or nails or other pieces of wood. They are transforming an industry from one that emits vast amounts of CO_2 during construction to one that actually stores CO_2 for the life of the building. It's not quite carbon positive, and it's so new that people are still arguing about how much carbon is actually stored or emitted, but it's a huge step in the right direction. We still have to worry about the state of our forests, and mass timber should only be made from sustainably harvested wood, and like any material, we should use as little of it as possible. However, use of natural, renewable materials is the best way to significantly reduce the upfront or embodied carbon emissions of buildings.

Operating Emissions from Buildings

Many in the industry talk about net zero as a good target for reducing carbon emissions. In this concept, one adds solar panels or other renewable sources that generate enough energy when the sun is shining and feed it into the grid to offset withdrawals from the grid at night or in winter. The trouble is that it doesn't really work because it is addressing supply rather than demand. On long summer days, there is often more power being generated than the utilities can use, and in many locations, the power given back in winter is "dirty," produced with coal or natural gas, so that even if it is net zero on an energy basis, it may not be on a carbon basis. The grid supplier has to maintain generating capacity to supply everyone in the depths of winter, which isn't very efficient, and it doesn't really scale vertically, where you can have lots of units on top of each other. Rooftop solar tends to disproportionately benefit people who have big roofs on single-family dwellings.

It's all so very status quo, like Elon Musk's dream of the big suburban house with a Tesla and a Powerwall battery in the garage and his solar shingles on the roof. This fantasy costs a lot of money, takes up a lot of space, has a huge embodied carbon footprint, and services a tiny overlap in the Venn diagram of rich people who say they care about the environment.

The Heat Pump Revolution

A decade ago, every time you read an article about green building, it was pushing geothermal heating, which in most cases wasn't geothermal at all; it was a ground source heat pump with the condensing end connected to pipes circulating a liquid in the ground. I was excoriated for refusing to call it geothermal (which should be reserved for systems that use heat directly from natural sources like hot springs, geysers, and volcanic hotspots). Others in the industry would call it solar heating, "a clean (no fossil fuel consumption) form of renewable energy that involves our sun heating the earth beneath our feet."[23] Except the pumps still run on electricity, and it has nothing to do with the sun; it is just moving heat from one place to another, like burying your refrigerator's coils in the ground.

You don't hear as much about geothermal heating and cooling any more because air source heat pumps are far cheaper and now do almost as good a job, without all the drilling and piping. So now the mantras are *electrify everything* with *air source heat pumps* (ASHPs).

ASHPs are truly the revolution of the last 20 years. They lose their effectiveness as it gets cooler and they can suck less energy out of the air; when they were first introduced, they could barely work below 0°C and had to be backed up with electric resistance heating. They still are far less efficient at very cold temperatures but now work down to −20°C.

The problem here is that electricity still costs three times as much per unit of heat than natural gas does, so it doesn't make sense unless you also reduce demand at the same time, which nobody seems to mention. So, people reduce the energy cost by getting big heat pumps, most of which have refrigerants that are serious greenhouse gases, as much as 1,700 times worse than CO_2. So again, you have to deal with the demand side first to reduce the size of the heat pumps and have more options for refrigerants. For example, you can buy small heat pumps that use propane as the refrigerant, called R-290. It's much more benign as a greenhouse gas (only 3 times as bad as CO_2), and it is used in small enough quantities that it is not considered a danger for fire or explosion.

If you just need heating, there are heat pumps that use CO_2, but they can only deliver hot water for heating. I am considering these for my own home after I invest in insulating our attic.

But in the end, we come around again to the same answer: you can't really electrify everything and push heat pumps as a sustainable option unless you reduce demand.

The Potential of Passivhaus

That brings us back to the Passive House or Passivhaus concept of reducing demand.

Passivhaus, as formalized by Wolfgang Feist and the Passivhaus Institute in Darmstatt, Germany, is defined by five design and construction principles:

- *Superinsulated envelope*: Like the Saskatchewan Conservation House, a complete wrap of what used to be considered ridiculous levels of insulation.
- *Airtight construction*: It's remarkable how much energy is lost just by leaking through walls.
- *High-performance glazing*: Often expensive imported triple-glazed windows.
- *Thermal bridge-free detailing*: Where heat goes through the walls at jogs, corners, bumps, and poorly designed details.
- *Heat recovery ventilation*: The building is so tight that you need controlled fresh air, an exhaust system that recovers the heat from outgoing air and heats the incoming air.[24]

The problem with all of this is that you can't see it: it is all in the walls, in the details and the design. There is a reason Thorstein Veblen came up with the term "conspicuous consumption" where consumers want to display wealth and income rather than deal with real needs. Veblen wrote that "in order to gain and hold the esteem of man it is not sufficient merely to possess wealth or power. The wealth or power must be put in evidence, for esteem is awarded only on evidence." It was called the Prius Effect, or Conspicuous Conservation, as noted in a 2012 study:

> The social signaling motive can distort private incentives and generate conservation investment that is individually rational but not social welfare maximizing. For instance, economists have begun to question whether homeowners over-invest in residential solar power because of its conspicuousness and under-invest in home insulation improvements, energy efficient heating and cooling systems, and window sealing because of the relative inconspicuousness of these investments.[25]

Passivhaus also makes life difficult for the architect because every jog and bump that they want to do to add interest to the building is an opportunity for a thermal bridge, and that big picture window with the great view gets really expensive. It takes talent to make a building "boxy but beautiful," as architect Bronwyn Barry calls it.

So, in the end, many people would rather spend money to increase the supply of energy with solar panels and fancy batteries rather than reduce demand with Passivhaus design. That way, they can have conspicuous conservation and big windows.

Passivhaus is all about inconspicuous conservation, which remains a hard sell. And even it doesn't go far enough.

The Passivhaus people have always been agnostic about how you get there and didn't consider embodied carbon when they created the standard. Designers have often used materials with high embodied energy, most notably plastic foam insulation. However, Emily Partridge, an architect with Architype in the UK, thinks we have to go past Passivhaus and proposes building with low-carbon materials, writing: "The embodied energy of new buildings can be reduced by using materials which use less energy to produce and are made from natural materials, such as timber and recycled newspaper insulation, instead of steel, concrete and plastic insulation."[26] This reduces the embodied carbon emissions, no need for a net here either.

This is the true modal shift in architecture and building, to switch to materials that get close to zero in upfront carbon emissions and have such low demand that they get close to zero in operating emissions. It is hard, and may not even be possible, given that electrical and mechanical systems are not made of wood, but it is a target worth striving for. It shows that it is possible to decouple our homes and buildings from fossil fuels and carbon, and should be a model for other sectors.

Form and Zoning

It helps if you make it smaller and simpler. It helps even more if you build multifamily dwellings with less outside wall exposure for each unit; this significantly reduces the operating emissions all year round. It's not about going net zero but about getting as close to zero as possible. Then if you want to get fancy, you can top it off with a little solar panel.

It's all very much like the 1.5-degree lifestyle target; we could do an incremental reduction every year, cutting emissions between now and 2030, or we can just decide to go bang and try it now. There is absolutely no reason that zero-carbon building couldn't be the standard tomorrow, except nobody is really taking this seriously; everyone still wants underground parking in their tall glass apartment buildings, and developers would rather sell you a granite countertop than a low-carbon energy-efficient condo. So instead, every new house and apartment and office building is going to make a big carbon burp as it's built and then lock in operating emissions for the next 50 years.

Sufficiency

Here again, there is the question of how much home do we actually need. Many people in Paris or New York live in small apartments because the city is an extension of their home. One reason modern homes have expanded so much is that we have basically privatized what the city traditionally did: the backyard replaced the park, the TV media room replaced the theatre and the movie house, the family room replaced the front porch or the stoop; everything got internalized. The double-wide fridge and the garage and the SUV's trips to Walmart and Costco replaced the bodega or the vegetable store or local butcher.

We will not be going back to an era where one bathroom on the second floor was enough for everyone; even in my own hundred-year-old house, the previous owners stuck a toilet and sink on the ground floor. My daughter won't even go into a movie theater; she is sensitive to perfumes, and you can watch everything online. But we can design homes efficiently with less space, smaller rooms, carefully designed multifunction spaces; we don't need 3,500 square feet for your average family of four when 950 was considered glorious in the 1950s—there is a happy Goldilocks floor area in there somewhere.

In her wonderful series of books that started with *The Not So Big House*,[27] Sarah Susanka showed how good design and planning made a huge difference in how much space you needed. She didn't suggest that it would save you any money, suggesting that "as a rule of thumb, a Not So Big House is approximately a third smaller than your original goal but about the same price as your original budget." That's how you pay for Passivhaus levels of efficiency, for healthy materials and high-quality finishes. "A not so big house feels more spacious than many of its oversized neighbours because it is space with substance, all of it in use every day." You don't have to give anything up to live in less space, you just have to design it better.

Of course, Sarah Susanka wrote all this before the COVID-19 pandemic, which may well encourage people to build bigger homes with more separate rooms, to adapt to the changing ways we work and learn.

The Third Industrial Revolution

I happen to have had a pretty good seat to witness the start of the Third Industrial Revolution, which we are living through now and was made possible by the computer and the shipping container. My father was a pioneer in the shipping container industry, building containers for the Canadian railroads back in the sixties and then founding one of the first container leasing companies; my first job was inspecting containers in his yard. His business partner in New York happened to own Commodore Business Machines, so I got to play with one of the first electronic calculators and Commodore 64 computers.

The container is to the Third Industrial Revolution what the train was to the second: an order of magnitude improvement in moving stuff at a fraction of the cost, so that the making of just about anything could move to where the labor supply is cheaper and other constraints, like pollution and safety controls, were not as strict. The computer and the Internet are changing the way we work as dramatically as the typewriter and the phone did in the Second Revolution.

In 1985, when I was an architect, I was profoundly influenced by an article in the *Harvard Business Review* by Philip Stone and Robert Luchetti titled "Your Office Is Where You Are."[28] The authors considered the implications of the first wireless office phone (it connected via infrared light) and how it would change everything about office design, that would "challenge the customary ways of thinking about offices and show how managers can gain the advantages and avoid the disadvantages of the new technologies." It seemed to me then and there that the traditional office was dead, that we were no longer tied down by wires to a fixed location.

But the office fought back, first with computers that tied us down with wires again, and with managers who wanted to see their employees' bums in seats. The office also fought back with "efficiencies" to make it more cost-effective, increasing the office densities, so private offices gave way to cubicles that gave way to basically shared desks.

The impact of teleworking was negligible among existing businesses. According to the Federal Reserve Bank,[29] prior to the pandemic only 2.5% of office workers were working from home full-time, or "telecommuting" as they used to call it.

But that was just among traditional firms. There were also many people starting up businesses, a whole new world of publications and ventures based on the internet, people who were not typical office workers. Other data suggest that as much as 11% worked from home. The tools, equipment, and programs that office workers needed to work from home all existed by 2020, ready to be put to work when the COVID-19 pandemic sent all the office workers home, when many like me could say "welcome to the revolution!"

Many of them are never going back. Consultant Kate Lister estimates that 56% of the American workforce is in a job where one could work from home, and that when it all shakes out, the "best estimate is that we will see 25–30% of the workforce working at home on a multiple-days-a-week basis by the end of 2021."[30]

Managers have found that they still can manage, that downtown real estate is expensive, and many workers are perfectly happy not to have to commute and are as or more productive than they were in the office.

Kate Lister also noted early in the pandemic that this is good for the environment:

> Even in the early days of the global response to COVID -19, we are starting to see a dramatic reduction in traffic, congestion, and pollution. While, sadly, sustainability has not been a primary driver of remote work in recent years, being able to actually see the difference it can make may finally flip the switch for employers and employees. The annual environmental im-

> pact of half-time remote work (for those who both want to work remotely and have a compatible job) would be the greenhouse gas equivalent of taking the entire NY State workforce off the road. The fact is, there is no easier, quicker, and cheaper way to reduce your carbon footprint than by reducing commuter travel.[31]

That's the Third Industrial Revolution kicking in, and it will change the way we live and work, and how we emit carbon.

The Rise of the 15-Minute City

The 15-Minute City is the timely new name for ideas that have been kicking around for decades. Jane Jacobs wrote in her last book, *Dark Age Ahead*, that a healthy neighbourhood is more than just a place for a nuclear family to sleep:

> Two parents, to say nothing of one, cannot possibly satisfy all the needs of a family-household. A community is needed as well, for raising children, and also to keep adults reasonably sane and cheerful. A community is a complex organism with complicated resources that grow gradually and organically.[32]

The New Urbanists address these ideas in their ten principles, two of which are:

> **Walkability:** Most needs are within a 10-minute walk of home and work. Street design is friendly to pedestrians, because buildings are close to the street and have porches, windows, and doors. Streets have lots of trees and on-street parking, with parking lots and garages placed behind buildings and houses, often connected to alleys. Streets are narrow, which slows traffic dramatically.
>
> **Mixed-Use and Diversity:** Neighborhoods, blocks, and buildings offer a mix of shops, offices, apartments, and homes. The neighborhoods welcome people of all ages, income levels, cultures, and races.[33]

When I was president of the Architectural Conservancy of Ontario, I pitched "heritage urbanism" as a way of learning from older buildings and neighborhoods about how to live a low-energy lifestyle:

"Heritage is about learning how we lived, how we built our houses and communities in the past, as a template for the future."

Before the pandemic hit, Professor Carlos Moreno of the Sorbonne updated the concepts and introduced the 15-Minute City, summarized by Natalie Whittle of the *Financial Times*:

> The concept of "la ville du quart d'heure" is one in which daily urban necessities are within a 15-minute reach on foot or by bike. Work, home, shops, entertainment, education and healthcare—in Moreno's vision, these should all be available within the same time a commuter might once have waited on a railway platform.[34]

It was picked up by Mayor Anne Hidalgo of Paris and is described by Feargus O'Sullivan in *CityLab* as "a commitment to bringing all life's essentials to each neighborhood means creating a more thoroughly integrated urban fabric, where stores mix with homes, bars mix with health centers, and schools with office buildings."[35]

Even that half of office workers who do not want to stay at home may not be going back downtown; many think that companies will set up satellite offices. Design firm HLW writes in the *Harvard Business Review* that a more distributed model is needed, to "better support employee performance and organizational resiliency while contributing to the improvement of the urban landscape and local communities."[36]

The other 44% of workers never will have the option of working from home: the teachers, the restaurant workers, the front-line health care professionals, the store clerks—they all have to go to where their patients and their customers are.

But their customers are no longer in the downtown office buildings; they are everywhere, in neighborhoods. They will still want to go out for lunch or go to the gym or buy a new pair of shoes. Many of the department stores folded during the pandemic; some of the malls are even being converted to Amazon distribution centres. Many of these needs may well be met locally, as shops and restaurants offering personal services return to our main streets.

This is not a new phenomenon. According to historian William Gribben, in 1822 the wealthy of New York City moved north to Greenwich Village to escape yellow fever, and the businesses that serviced their needs went with them.

> Along with the population, the infrastructure supporting the wealth of the city was also temporarily displaced, with the custom house, post office, and other offices moving up to Greenwich Village. Relocated financial institutions clustered on Bank Street, which still bears that name today.[37]

Even our health care services may decentralize; my doctor is now part of a family health team that is part of a much larger downtown hospital network, designed "to give you the very best primary care, when you need it, as close to home as possible." It may well become common. Andrew Nikiforuk wrote in the *Tyee* about lessons from the pandemic in Italy:

> To avoid a collapse of hospital systems the doctors propose that Italy and other nations quickly develop facilities in the community such as home care and mobile clinics to treat less severe patients.... The only way to prevent a similar disaster in other countries is to begin a massive deployment of outreach services that keeps as many patients as possible in their homes or other community-based settings.[38]

The Second Industrial Revolution ran on coal and then on gasoline; many predict that the Third will run on solar and wind and renewables. But that is still thinking about the economy continuing as it is, perhaps with people driving electric cars from net-zero houses in the suburbs to efficient solar-powered offices, basically a low-carbon upgrade of the way we live now. It's just a continuation of business as usual, the same old economic system designed around "extracting, processing and transforming energy as resources into energy embodied in products and services." We don't have the time or the money or the materials for that.

Instead, we have to think about how and where we live, where we work, and how we get around. We have to think about *modal shifts* from detached to duplexed to apartment homes, from cars to e-bikes to bikes to walking, about sufficiency, about what we actually need to get the job done well.

I live in the 15-Minute City, and it works. I can walk or bike to dozens of restaurants, four major grocery stores, and who knows how many bakeries, coffee shops, and gelati outlets. I live in a neighbourhood where you really don't need a car.

And, as I used to say in the heritage biz, it's not a relic from the past, it's a template for the future.

What Do We Learn from the Spreadsheet?

Not much, because there are five people in two families living in the house in two apartments, and it is really hard to separate out individual actions, but over half of those basic operating emissions are from burning gas. To really make a difference in household emissions, I would have to make a major capital investment in heat pumps. I do track personal water use; a shower is half the footprint of a bath. If I had not renovated the house into two apartments, this project would be almost impossible; my emissions per capita would be double. But even so, there is just that ongoing daily basic operating carbon emissions plus a shower are almost exactly the entire 2050 target of 1 tonne of CO_2 per year, or 2,739 grams per day.

My inability to actually measure anything or move the needle here shows how intractable and difficult the housing sector is. Living in a multifamily house with everyone on different schedules confuses smart thermostats. Turning out all the lights makes almost no difference when they are all LEDs already, so that my individual actions are almost meaningless. You can do all those things that every green guide says, from unplugging power supplies or dimming your monitors, and frankly and honestly, it would make almost no difference. I have done all the caulking and sealing that I can do, but it is nibbling around the edges. It is the big things, the heating and cooling, that

Table 6.1. Calculations for housing domain

	Item	Source	Unit	CO_2/ unit	estimates/ day	Nov 9	Nov 25	Nov 26	Nov 27	Nov 28	Nov 29
Housing											
Water heating	Shower	berners lee	each		600.00		600				600
	Bath	berners lee	each		1,100.00			1,100	1,100		
	Basics	fixed operating (gas, electric, water			2,001.20	2,001	2,001	2,001	2,001	2,001	2,001
Gas cooking	special		grams/hr		245.00						

really matter, and I cannot change those without a very big capital investment in insulation and heat pumps.

In the meantime, I assuage my guilt by using Bullfrog electricity and gas, where I pay extra to have renewables injected into the supply. I am paying for some wind and water power, and evidently to buy some methane from a garbage dump near Montreal. But offsets can't save us, and I do not include them in my calculations.

What Can We Do?

- In new buildings, natural materials with low embodied carbon combined with super-insulation, high-quality windows, and careful sealing to Passive House standards are really the only way we should be building.
- In existing buildings, reduce demand as much as possible by sealing, insulating attics and basements and affordable, accessible areas, and meet that reduced demand with air source heat pumps.
- In terms of your carbon footprint, apartments are better than duplexes or townhouses, which are better than single-family houses.
- 15-minute cities, transit-oriented developments, walkable communities, bicycle lane networks all contribute to lower-carbon living.
- Check out what you can walk to in 30 minutes, not an unusual time for a commute. Even in the suburbs, it will include a surprising variety of places to shop. Try that instead of driving.
- Electrify everything.

– 7 –

How We Move

From the 1.5-degree lifestyle report: "use of owned transport equipment and transportation services for commuting, leisure, and other personal purposes, e.g., cars, motorbikes, public transport, air travel, bicycles." Emissions from business purpose trips are included under respective domains of the products or services supplied.

Estimated carbon footprint of building a Tesla Model 3:

- Tesla Model 3 manufacture excluding battery: 7 tonnes
- Tesla battery: 5.750 tonnes
- Total: 12.750 tonnes of embodied carbon.

There are many different visions of how we should be living and moving in a low-carbon world. Elon Musk has an interesting one that he is working hard to promote. Here's my description of his vision:

> We will live in beautiful suburban houses on wide lots with two-car garages filled with a Tesla SUV and a smaller Model 3. They will be powered by Tesla Powerwall batteries ($7,600 each) charged by Tesla's solar shingles (roughly US$22 per square foot) mounted on the big south-facing roof. In 2016, Musk got stuck in traffic and tweeted, "Traffic is driving me nuts. Am going to build a tunnel boring machine and just start digging." And, so, he started the Boring Company to drill tunnels for his cars to drive through, under the traffic. Since parking will be a problem, the car will be self-driving and will return home on its own until it is required.

Some of the problems with this vision include that it is really expensive and will be available to a very small segment of the population, or that it is sprawl-inducing because everyone needs that big roof to charge those big batteries. But perhaps the biggest one is that if the goal is to reduce our carbon footprints, it still has to address the fundamental issue of the embodied carbon, the upfront carbon emissions from making all this stuff.

I will keep repeating that "when you start looking at the world through an upfront carbon lens rather than operating carbon, everything changes." We have to look at numbers that are a lot bigger than my chicken takeout dinner. The most straightforward example is the electric car, in this case a Tesla Model 3. The traditional way of looking at electric cars vs internal combustion engine (ICE) powered cars is with a life cycle analysis, where one totals the upfront carbon emissions (UCE), the operating emissions over the life of the car, and the end-of-life emissions from getting rid of the car. It used to be thought that the UCE of electrics were twice that of ICE cars because of the battery manufacturing, but it's getting greener and more efficient every day. The most recent study, from Auke Hoekstra of the Eindhoven University of Technology, compares the lifetime greenhouse gas emissions of gasoline and electric cars and shows how far electric cars have come.[1]

When you look at the savings in carbon from operations (even in parts of the world where the electricity was made from coal), that extra carbon debt is paid off after about 26,000 km, or in about two years. Over the estimated life of the car, the carbon savings are significant, but they still have lifetime carbon emissions of about half those of ICE-powered cars because of the upfront emissions. That is still simply too high, although it may be an underestimate. We don't even know the estimated life of an electric car; they haven't been around long enough. Hoekstra does his calculations based on a 250,000 km useful life, but electric cars have far fewer moving parts and require far less maintenance, the batteries are not deteriorating as quickly as it was thought, and they might last twice as long as ICE-powered cars. We won't know the full life cycle emissions until we have been through a full life cycle. The problem is, we don't have a life cycle. Even if a Tesla Model 3 lasts ten years, that's still 2.2 tonnes per year, which doesn't leave much left over for anything else. This is the fundamental reason that we can't plan to simply replace all our ICE cars with electric cars; instead, we have to plan to have fewer cars.

It's not just the upfront emissions that will be a problem in a rapid switch to electric cars. In his new book, *Less Is More: How Degrowth Will Save The World*, Jason Hickel describes the problems with the materials that go into them:

> Lithium is another ecological disaster. It takes 500,000 gallons of water to produce a single ton of lithium. Even at present levels of extraction this is causing real problems. In the Andes, where most of the world's lithium is located, mining companies are burning through the water tables and leaving farmers with nothing to irrigate their crops.[2]

He writes about scientists' concerns about other materials:

> They agree, of course, that we need to end the sale and use of combustion engines and switch to electric vehicles as quickly as possible. But they pointed out that replacing the world's projected fleet of 2 billion vehicles is going to require an explosive increase in mining: global annual extraction of neodymium and dysprosium will go up by another 70%, annual extraction of copper will more than double, and cobalt will need to increase by a factor of almost four—all for the entire period between now and 2050. We need to switch to electric cars, yes; but ultimately, we need to radically reduce the number of cars we use.[3]

There is no question that batteries will continue to get more efficient and that many of these rare elements are being designed out of them; Tesla is already using cobalt-free batteries,[4] and the amount of lithium needed is decreasing. However, we are seeing the same old thing that we saw with SUVs: as they got more efficient, they got bigger and heavier. Soon we will have electric F-150s and Rivians and Cybertrucks and even electric Hummers, a whole fleet of giant vehicles with giant battery packs. Hoekstra estimates emissions from battery production to be 75 kg of CO_2 per kWh (on the low side of the estimates out there). The Rivian's biggest battery pack is 180 kWh, which packs a carbon footprint of 13,500 kg, or 5.4 times my annual carbon budget, just for the batteries.

Others have said that the grid cannot support all those extra electric cars, that it might even have to double in size. In fact, their batteries could actually become part of the grid, storing renewable power and delivering it back when needed, smoothing the grid. This is not going to be the deal breaker.

The real insurmountable problem is the upfront carbon emissions. Suddenly, the idea of selling the family Subaru and buying a Tesla doesn't seem to be such a win for my carbon footprint; whether it's 7 tonnes of embodied carbon for the ICE car or 10 for the Tesla, or as much as 60 for the Hummer EV, it still blows the carbon budget. Hoekstra also calculates that it takes "only" 30,000 km of driving to pay back the extra footprint of the battery, but the less you drive, the longer it takes. If you drive as little as we do, that still takes five years.

In my own case, I just don't drive at all in the city, but I have not been able to convince my wife to do the same, so we still have a family car, a four-year-old Impreza. The plan is to drive it as little as possible and count every litre of fuel, and to drive it into the ground.

But even a bit of driving can blow your budget. I spend the summer in a cabin 250 kilometers north of Toronto. I go up in June and come back in September, and probably do most of my year's driving just taking the garbage to the dump. But recently I had to drive to the closest town with an Apple repair depot to fix my keyboard, a 76-kilometer round trip, not much further than lots of people do on a daily commute. That turned out to produce 12.160 kg of CO_2, equivalent of two days of my carbon budget, and that is just based on the direct consumption of gasoline. Had I used Hoekstra's estimate of lifecycle emissions per km for a comparable gasoline-powered car, that little trip would have emitted 19.760 kg of CO_2. Furthermore, we are only counting the emissions from actually burning the fuel, not producing it in the first place; for Canadian oil, that probably increases the emissions per km by 50%.[5]

Cars are wonderfully convenient and make my summer lifestyle possible, but they are climate killers. Basically, driving anywhere is inconsistent with a 1.5-degree lifestyle. My Subaru lasted 16 years, and my intent is to never buy another. In an aside, Subaru and Toyota, two companies that have painted themselves as green, supported the Trump administration's rollback of emission standards; Ford and Volkswagen do not. I would not be buying a Subaru today. Having learned what I have doing this exercise, I would not be buying any new car at all.

The Electric Car Is Still a Car

The other problem with the electric car is that it is still a car, a product that thrives in a suburban low-density ecosystem. Cities adapted to it by almost destroying themselves, razing buildings for parking, and pushing pedestrians and pushcarts off the streets, narrowing sidewalks and adding lanes. Lewis Mumford wrote in his book *The Highway and the City* in 1964: "The right to have access to every building in the city by private motorcar in an age when everyone possesses such a vehicle is the right to destroy the city."[6]

According to MIT planning professor Eran Ben-Joseph, there are 500 million parking spaces in the US, covered in asphalt or concrete.[7] Bridges continue to be built and roads continue to be widened, all made with steel and concrete. Elon Musk is not happy about getting stuck in traffic, so he is now drilling concrete tunnels for his electric cars. The carbon cost of all that infrastructure has been estimated by Hall and Klitgaard (and quoted by Bart Hawkins Kreps) to be "about 38% of the energy used by fuel itself,"[8] so if an average car puts out 200 grams of CO_2 per kilometer, the carbon footprint of the infrastructure is about 76 grams per kilometer, which would apply whether it is electric or gasoline.

As noted before, "land use and transportation are the same thing described in different languages"—cars *are* the suburbs. So, one can't just say (as many do) that electric cars will save us. They are incompatible with the development patterns that we need to promote; the infrastructure they need to be useful is too carbon-intensive.

Not long ago, I got in an argument with a leader in the "electrify everything" crowd who got angry and dismissive with me and others who were talking about the role of the car. He claimed that we were going to alienate the vast majority of the public, that it was one thing to try and convince them of the need to fix their houses and get heat pumps and get rid of gas, but that it was counterproductive and stupid to talk about getting rid of cars. Perhaps so, but ultimately, we do not have a whole lot of choice if we are serious about dealing with this crisis.

Moving to the Country

When you know all this, you start thinking differently about how we get around, and about how we live. Because of the coronavirus, all kinds of people are thinking about moving to the country, but as I have learned, this has its own issues. When I am at my little cabin in the woods, not using any air conditioning, not going anywhere, and living a really low-carbon lifestyle—just a little propane for the barbecue and a few liters of gas per season for the outboard motor—the fact is that I am dependent on the car for everything. The dump is 3 kilometers away. The nearest store, where we get our basics and even our drinking water, is 5 kilometers. I was thinking of bringing my e-bike up for the summer, but I cannot carry a jug of water on my bike.

Everyone up here drives big pickup trucks; they don't want to pay the prices in the local store, so they drive 40 kilometers for every grocery run to the bigger towns. Their houses are not well insulated, and there is no gas, so they heat with expensive electricity (subsidized by the provincial government, these are conservative voters) and burn a lot of wood. Some would say that's fine, the electricity is hydro powered and the wood is carbon neutral, but it's not fine; they are working to pay their energy bill and spending their spare time chainsawing trees and cutting, splitting, and stacking firewood. Burning the wood emits piles of particulate matter smaller than 2.5 micrometers (PM2.5) that, even at the low population densities, is still problematic for people's health. Yet when I go into the local hardware store and buy a bag of firewood, the label actually says that it is imported from Norway; our economic system is so distorted that it is easier for the store to just order a bag of Norwegian wood than it is to make a deal with a neighbor to supply the local stuff, which I now do.

Two and a half hours north of Canada's largest city, the year-round population is incredibly tiny, thousands of square kilometers with fewer inhabitants than a couple of apartment buildings in the city. Outside of a few who chop trees for the paper mills, everyone survives by servicing the needs of the rich people who drive their Porsche

Panameras and Land Rovers from the city and need a new boathouse, or the less rich people like me who need the wasp nest removed from my outboard motor. The entire economy runs on vast quantities of petroleum needed to move everything and everyone around. I may be self-righteous and proud rowing my Hudson single scull around the lake every morning and not owning a big wakeboard boat or a Sea-Doo, but the five days' worth of carbon that it takes me to get here in my little Subaru makes such arguments specious.

The E-Bike Revolution

When I am in the city, I now get everywhere on an e-bike. I have been test-driving them for a few years, but never thought I needed one, given the kind of riding that I did. But when the Dutch bike company Gazelle offered to let me try one out, I fell in love and bought it at the end of the test. I ride this e-bike farther and more often than I ever did on my bike, and research shows that my experience is not unusual.

The carbon footprint of my Gazelle e-bike is not very big. A rough calculation based on 22 kg of European steel gives the bike a footprint of 45 kg of CO_2e, while the battery, at 75 kg/kWh comes in at 37 kg, totalling 82 kg, about 1/146 that of a Tesla Model 3.

That's still almost two weeks of carbon, and I already had a bike. Conversion kits and even electric add-ons like the Copenhagen Wheel could have reduced my footprint, but now I have a safe, solid bike that will run for years. If it keeps me out of a car (and it did all last year), I will take the carbon hit.

A recent article, "The E-Bike Potential: Estimating Regional E-Bike Impacts on Greenhouse Gas Emissions" included some interesting data on the carbon footprint of different modes of transport:

> The European Cyclists' Federation found that bicycles emit 21 g, e-bikes emit 22 g, buses emit 101 g, cars emit 271 g lifecycle CO_2e per person kilometer in Europe. Additionally, according to the European lifecycle emission estimates of Weiss et al. (2015), bicycles emit 5 g, e-bikes emit 25 g, buses emit 110 g, and cars emit 240 g CO_2e per person kilometer. Clearly, e-bikes emit little more than conventional bicycles and far less than cars and buses, even when considering manufacturing, use, and disposal.[9]

The European Cyclist foundation data seem sketchy: an e-bike has a battery and needs charging, and it seems more likely that the larger difference with Weiss is a better reflection of reality. Either way, my e-bike emits between a tenth and a twelfth that of a car.

The wonder of the e-bike is that it so radically expands what two wheels can do. It opens cycling to older people, those with disabilities, people who live in hilly cities where regular cycling requires serious effort. It flattens hills and distance. A former co-worker has cystic fibrosis and now just throws her oxygen tank in the carrier and is cycling around Atlanta. It flattens seasons too; you dress as you would for a walk, knowing that you will not work up a sweat if you don't want to.

The previously mentioned article demonstrated that if just 15% of a city's population switched to e-bikes, it would reduce the carbon emissions from transportation by 12%; 15% is nothing; in Copenhagen 50% of people ride. Fifteen percent is also not a stretch at all, and a higher percentage is possible, but not if you only talk about the bikes themselves; they have to be part of a bigger package.

3 Things Are Needed for the E-Bike Revolution

Decent Affordable E-Bikes

Whereas e-bikes have been popular in continental Europe for years, they are just beginning to have a significant impact in North America. Since bikes were seen more as recreation than transportation, e-bikes were seen as "cheating"—you are not getting as much exercise. They were often conflated with electric scooters, the Vespa-like things with dinky useless pedals, that were often being driven by people who lost their licenses for DUI.

Then there was the patchwork of regulations across North America, confusion about whether e-bikes are bikes or some other form of vehicle. This was all figured out in Europe years ago, where Pedelec e-bikes, which had 250-watt motors and no throttle (but picked up the riders' pedalling and gave them a boost) and a top speed of 20 km/hr were treated like bikes.

American exceptionalism being what it is (More hills! Longer distances! Faster traffic! Heavier people!), they had to reinvent the wheel and have a 750-watt maximum, 28 km/hour limit, and throttles so riders can just sit there like on a motorcycle, instead of being on a bike with a boost. But at least there were now rules, and companies like Rad Power Bikes started selling decent e-bikes for under $1,000 (my Dutch-built Gazelle costs three times that). They sell them online, which I originally thought was a terrible idea, thinking we should support our local bike shops and make sure that they are assembled properly by experts, but many people, mostly women, told me that so many bike shops are staffed by misogynist bike snobs who treat e-bike shoppers dreadfully. They convinced me that buying online wasn't such a terrible idea.

A Safe Place to Ride

Since most politicians and planners considered bikes to be recreational, they were loath to give up any road space for bike lanes, and every one of them became a contentious political battle. Most North American bike networks are patchy, inconsistent, and full of parked cars because they are not properly separated.

When the pandemic hit, many cities suddenly became big fans of bike lanes, given the dramatic increase in ridership due to people wanting to avoid public transit. It's hard to tell how many of these lanes will remain after the bug is gone, but I suspect that many people who took to bikes and e-bikes out of necessity will fall in love with them.

But for bike lanes to work, the network has to be continuous, not just dumping you into the middle of a busy street. It should be protected, so that it doesn't become the FedEx lane. It needs to be maintained and properly ploughed. In Copenhagen, they clear the lanes before they do the streets. They have to be treated like proper road infrastructure, not as an afterthought.

A Secure Place to Park

Parking remains the missing link. Whereas zoning bylaws have required car parking for decades, they are just beginning to require bike parking. Municipal facilities are few. Systems being proposed in North America include Shabazz Stuart's Oonee, an interesting modular system of bike storage lockers that is advertiser supported. But he is having trouble finding places to put them and is getting little municipal support. We have such a long way to go in all three of these issues. I follow Shabazz Stuart's Twitter account from New York City; he tweeted in August, 2020:

> Sad story to share @NYC_DOT. Was at local bike shop when a young woman showed up to donate her bike. She was throwing in the towel. Had been excited to #bikenyc to work but got doored by a taxi (she was ok) then had her seat stolen. So she's done. We failed her. Do better.[10]

We all have to do better. In the Netherlands or Copenhagen, vast multilevel secure bike parking lots at train and bus stations encourage multi-modal transport; in the cities, bike parking is everywhere. This is going to be needed in North American cities too for e-bikes to really take off as a form of transportation.

And it is going to take off, because people are finding that e-bikes are effective transportation alternatives. A recent study found that people who switch to e-bikes increased their travel distance from 2.1 to 9.2 km per day on average, and the use of the e-bike as a share of their transportation increased from 17% to 49%.[11] That is a serious *modal shift*.

When Everything Is in Place, It Can Make a Huge Difference in Your Transportation Footprint

In this book, we are sticking to the personal, so let's look at what my e-bike does for me. The city of Toronto where I live is built on the north shore of Lake Ontario, and most of the city is built on a tilt, all sloping down toward the lake. A few miles north of the lake, there is a steep escarpment, the old shoreline left over from the last Ice Age when the lake was much bigger. On a regular bike, riding down to work or school was always a breeze, but at the end of the day, you had a long slog through the sloping city, with a really big hill right at the end. The e-bike flattens out the city, and the escarpment is no longer daunting.

I find now that I am always on the bike, pretty much year-round (last year there was one day in winter when I didn't ride to teach, the snow had not yet been cleared). Twenty-five grams of carbon per kilometer? I can live with that.

When you ride an e-bike, hills don't matter. Weather matters, but not as much as when you are riding a regular bike because you don't need to work up a sweat, so you just dress as if you were walking. Snow matters, but that is a governance problem of taking bike lane clearing seriously, which they do in Scandinavia but not yet in North America.

All of this leads me to conclude that e-bikes are a far better way of dealing with transportation emissions than electric cars. They won't work for everyone, but they don't have to. Imagine if we gave a fraction of the attention to bike and e-bike infrastructure and subsidies that we do to automobiles, it could change everything.

The Problem with Flying

Business trip, Toronto to New York City, March 2020:
Lift to UP Express: 656 grams of carbon
UP Express station to airport: 651 g
Flight to NYC: 90,000 g
Taxi to Times Square: 6,900 g
Limo to La Guardia: 13,800 g (giant Chevy Suburban)
Flight to Toronto: 90,000 g
UP Express and subway home: 682 g
Total for trip: 202,689 g (202.68 kg!)

After I started tracking my carbon footprint for this project in January 2020, everything was going very nicely until the website I write for was bought by a big American company, which asked us all to come to New York to meet the new boss. It's hard to say "Sorry, I am on a low-carbon diet" to that, so I got on the plane from Toronto to New York City, and took a cab or two because I was in a hurry. In the 36-hour trip, I burned through 202.68 kg of carbon, equivalent to 31.2 days of my carbon ration. For anyone seriously trying to live a low-carbon lifestyle, you have to stay on the ground.

On the other hand, in the larger scheme of things, many are saying we make too much of flying. They point out that aviation is only 2% of worldwide carbon emissions, although this is disputed. Parke Wilde of Tufts University[12] (who is trying to get academics to stop flying) says it's really 2.97%, which is still less than C40 Cities says is the impact of our clothing. He also reminds us of "radiative forcing," the additional impact of the emissions at high altitude, that bring it up to roughly 4.9%.

A new article, "The Contribution of Global Aviation to Anthropogenic Climate Forcing for 2000 to 2018," examines radiative forcing and finds non-CO_2 effects to be three times as bad as the CO_2 itself, which would bring the total up to 8%.[13]

Aviation is unusual in that it is international, so nations didn't have to include its emissions in their national totals under the Paris Accords, so there is no real incentive to cut back. Notwithstanding the hydrogen hype that Airbus has been pushing out, there are not a lot of alternatives to petroleum-based jet fuels. If it gets back to its former growth rate after the pandemic, while everything else gets greener in an attempt to hit the 1.5-degree targets, then aviation could actually dominate carbon emissions: Carbon Brief estimates that aviation could consume as much as 27% of the carbon budget by 2050.[14]

None of this includes the emissions from what the *Economist* calls the Airline Industrial Complex that supports the 4.5 billion passengers and 100,000 flights per day we saw in 2019, before the pandemic: "These journeys supported 10m jobs directly, according to the Air Transport Action Group, a trade body: 6m at airports, including staff of shops and cafés, luggage handlers, cooks of in-flight meals and the like; 2.7m airline workers; and 1.2m people in planemaking."[15] That's a lot of people making a lot of stuff. As Parke Wilde notes, we have to consider the "transportation to the airport, the energy emissions to produce and transport jet fuel, the ground operations for airports, and the embedded emissions for everything from the aircraft themselves to the airport infrastructure."[16]

A few silly back-of-the envelope calculations show the scale of this: a 737 weighs 41,000 kg, mostly virgin aluminum and magnesium, with a footprint of around 450 tonnes of CO_2 just for the metal. The builders of the just-completed Beijing Daxing Airport proudly announced how much steel and concrete went into its construction: it has a footprint of 656,000 tonnes.[17] There are 23,600 planes in commercial service in 2017,[18] and prior to the pandemic, this was expected to double by 2030.

The only reason aviation is such a low percentage of global emissions is that so few people do it; 80% of the world's population has never been on a plane. Only half of Americans fly in an average year.

Until the pandemic hit, aviation was growing every year; in Indonesia, planes replace ferries. The Chinese became the world's most numerous tourists. Flying kept getting cheaper, and more people were doing it. It would have not been long before we realized that flying and all of the associated carbon footprints were a lot bigger than blue jeans.

It's one of the toughest nuts to crack in the 1.5-degree lifestyle. I miss not going to Berlin or Portugal, sites of Passivhaus conferences where I was invited in 2020, both cancelled by the pandemic. I love travel, I even love airports, and I love these conferences. But my 2019 flying footprint was 6.85 tonnes, and I am not alone; as noted earlier, the richest 10% of the world's population—me included—emit 43% of the carbon, and they do all the flying.

But as Parke Wilde of FlyingLess[19] keeps saying, the conference circuit has a huge footprint; "Researchers at the University of British Columbia[20] recently estimated that air travel emissions are equivalent to 63–72% of emissions from campus operations." Some justify this. Andrew Waugh, principal of Waugh Thistleton Architects and the world leader in building and promoting wood construction, is satisfied that every flight he takes to a meeting or a lecture and convinces someone that their concrete or steel building should be wood, he has more than compensated for the footprint of the flying. I suppose if I convince a few more people to build super-efficient Passivhaus buildings (out of wood!), then I could say the same thing. Climate scientist Michael Mann uses a similar justification:

> Though air travel accounts for only a paltry 2% of global emissions, whether or not climate scientists should fly consumes far more than 2% of my Twitter timeline. Unfortunately, sometimes doing science means traveling great distances, and we don't always have the time or luxury to take slower low-carbon options. We have a job to do, after all.[21]

However, there are many climate activists and scientists who have given up flying, recognizing the huge role in their personal footprints.

Scientist and author Peter Kalmus did and believes that it sets an example, writing in *Grist*: "By changing ourselves in more than merely incremental ways, I believe we contribute to opening social and political space for large-scale change. We tell a new story by changing how we live."[22]

Conferences are probably dead for a while, thanks to the pandemic, so I don't have to worry for now. People are also learning how do a better job producing virtual conferences, which are also open to a much larger and diverse audience. It may well be that conferences stay virtual, and that we all tell a new story by changing how we meet and communicate.

What Do We Learn from the Spreadsheet?

The very obvious conclusion is that driving a gasoline-powered car is completely incompatible with living a 1.5-degree lifestyle; one trip up north is three days of my carbon budget. Just driving a 10 km round trip to buy drinking water is a quarter of it. And don't even think of flying.

But it was also a bad time to try this experiment, given that during the pandemic I really didn't go anywhere. I was not on public transit once and barely used my e-bike. Working from home as I do, with grown kids, I am not really a representative sample.

What Can We Do?

- Get out of your car; it is the biggest single factor in most people's footprints. Note I didn't say get rid of your car; we still have one for longer trips and occasional use. But use it as little as possible.
- Choose where you live carefully; it is the single biggest influence

Table 7.1. Calculations for mobility domain

	Item	Source	Unit	CO_2/ unit	Estimates/ day	Nov 25	Nov 26	Nov 27	Nov 28	Nov 29	Nov 30
Mobility	e-bike	calculated	grams/km	17				136			
	streetcar/ subway	OWID	grams/km	31	600.00						
	bus				1,100.00						
	Subaru	Terrapass	grams/km	160	2,001.20		800				
	Aviation										
	outboard on boat		grams/km	141.83	245.00						

on how much you drive. Where you live defines how you get around.

- In order of impact: walk, bike, e-bike; and many now are looking at e-cargobikes.
- Fly as little as possible.

– 8 –

Why We Buy

From the 1.5-degree lifestyle report: "Consumer goods: goods and materials purchased by households for personal use not covered by other domains, e.g., home appliances, clothes, furniture, daily consumer goods."

Why Do We Consume?

The 1.5-degree study specifically considers consumer goods as those things not covered in, say, nutrition or mobility or housing. But can they really be separated? A car is a consumer good, as is a home. The choices and decisions we make about them are made in much the same way as the choices we make about furniture or clothing. The home appliance like a stove or fridge is a tool for nutrition, and the car, a tool for mobility. So, for almost every category, and not just consumer goods, we have to figure out why we buy, what motivates us.

Recall what economist and physicist Robert Ayres wrote: "The economic system is essentially a system for extracting, processing and transforming energy as resources into energy embodied in products and services—*the purpose of the economy is to turn energy into stuff* [emphasis added]." But there is no point in making stuff unless someone is going to buy it. The stuff has gotta move. In his book *The Waste Makers*, Vance Packard quotes banker Paul Mazur:

> The giant of mass production can be maintained at the peak of its strength only when its voracious appetite can be fully and continuously satisfied. It is absolutely necessary that the products that roll from the assembly lines of mass production be consumed at an equally rapid rate and not be accumulated in inventories.[1]

Packard also quotes marketing consultant Victor Lebow:

> Our enormously productive economy...demands that we make consumption our way of life, that we convert the buying and use of goods into rituals, that we seek our spiritual satisfactions, our ego satisfactions, in consumption.... We need things consumed, burned up, worn out, replaced, and discarded at an ever-increasing rate.[2]

This is why the car-dominated suburban lifestyle was such a success at creating a booming economy in North America. So much more room for stuff, for consumption, creating a need for endless consumption of vehicles and the fuel to power them and the roads to run them on. For the hospitals and the police and all the other parts of the system. It would be hard to imagine a system that turns more energy into stuff. It is why houses get bigger and cars turn into SUVs and pickup trucks: more metal, more gas, more stuff. It is why governments are loath to invest in public transit or alternatives to cars: a streetcar lasts 30 years and doesn't add to consumption of stuff; there is nothing in it for them. They want a booming economy and that means growth, cars, fuel, development, making stuff. It's why they build tunnels in Seattle and bury streetcars in Toronto and fight over parking in New York City: Rule 1 is never inconveniencing the drivers of cars; they are engines of consumption.

For years, going back to the 1930s, there has been talk about planned obsolescence being built into products. One industrial designer told Vance Packard:

> Our whole economy is based on planned obsolescence, and everybody who can read without moving his lips should know it by now. We make good products, we induce people to buy them, and then next year we deliberately introduce something that will make those products old fashioned, out of date, obsolete.... It isn't organized waste. It's a sound contribution to the American economy.[3]

Packard thought it was more complex than that, and that there were different kinds of obsolescence. They are important distinctions:

Obsolescence of function. In this situation, an existing product becomes outmoded when a product is introduced that performs the function better. We have been overwhelmed with this in the computer and iPhone age, where the technology really has evolved so rapidly. My recent iPhone 11 Pro has finally supplanted my digital camera, and is demonstrably better than the iPhone 7 it replaced. Before that, I used my 35 mm Olympus camera for 25 years, and then went through six digital cameras in the last 20 years before giving up on them altogether.

But I would make the case that obsolescence of function is a good thing; look at what that phone can do. If you went to Radio Shack in 1990, you would have to buy fifteen different pieces of equipment, from telephones to answering machines to camcorders to calculators; Buffalo writer Steve Cichon calculated that it would cost $5,100 in today's dollars, and it would fill a room.[4] Now that the iPhone camera is so spectacular, I cannot imagine buying another separate camera. When it comes to obsolescence of function, I say bring it on.

Obsolescence of quality. "Here, when it is planned, a product breaks down or wears out at a given time, usually not too distant."[5] Packard quotes a marketer in 1958: "Manufacturers have downgraded quality and upgraded complexity. The poor consumer is going crazy."[6] Imagine if he were around today. When we renovated our house, I bought a lot of Samsung appliances, all of which have electronic controls. The kitchen stove lasted a year before it wouldn't hold the temperature; each repair cost hundreds of dollars as the entire electronic guts had to be replaced. I feel incredibly guilty admitting that I gave up and replaced the stove after four years; my mom's electric range lasted for ten times that long. I bought its replacement on the basis of reviews of dependability and ease of maintenance.

Obsolescence of desirability. "In this situation a product that is still sound in terms of quality or performance becomes 'worn out' in our minds because a styling or other change makes it seem less desirable."[7] This is what drives fashion and interior design; the minimalist look is the thing one year, the cozy hygge look the next. Or people ripping out their granite counters to replace them with more fashionable quartz. Or why every year the fronts of pickup trucks get more aggressive and dangerous looking; drivers of older trucks look like wimps. As Paul Mazur noted, "Style can destroy completely the value of possessions even while their utility remains unimpaired."[8]

This is what is really driving so much waste, what has also been called psychological obsolescence. It's why buying used or buying classic makes so much more sense, buying design instead of style. I am writing this at a desk designed by George Nelson in 1954 that is still a functional classic. Nelson himself said in *Industrial Design Magazine*:

> Design...is an attempt to make a contribution through change. When no contribution is made or can be made, the only process available for giving the illusion of change is "styling." In a society so totally committed to change as our own, the illusion must be provided for the customers if the reality is not available.[9]

We have to learn to consciously differentiate design from style.

The Linear Economy Is Based on Waste

A continuing challenge in writing this book is hewing to the categories set out in the 1.5-degree lifestyles report because everything is so interconnected; that's why I wrote that "land use, transportation, and energy are the same thing described in different language." One could add plastic waste to the list: 42% of all plastic (146 million tonnes per year[10]) is used for single-use packaging. In 2015, only 19.5% was recycled, 25.5% was incinerated, and the majority, 55%, was "discarded"—littered or landfilled.

In 1950, only 2 million tonnes of plastic were manufactured in the entire world, and none of it was for single-use packaging. By 2015, total plastic production was up to 381 million tonnes. After cigarette butts, according to UNEP, the biggest sources of plastic waste and pollution are "drink bottles, bottle caps, food wrappers, grocery bags, drink lids, straws and stirrers."[11] Much of this is related to the takeout food industry, which creates the category challenge: should I be writing this under food? Or transportation? Or housing? Or, finally, should it be energy because plastics are essentially solid fossil fuels? It's really all of the above.

According to Emelyn Rude, among the first specifically designed disposable take-out containers were "oyster pails" developed when oysters were cheap and incredibly plentiful on the east coast of the US.[12] Oysters are difficult and dangerous to shuck (I still have a scar on my palm from prying oysters when I was 13.) so the little cardboard box with the wire handle was developed to let people take them home pre-shucked. In the 1920s, a Chinese restaurant in Los Angeles started using them for take-out food, and they caught on fast, an easy way for people to eat exotic at home.

But if you wanted a coffee or a sandwich, you went to a diner. It might be a stand-alone, or it might be in a Woolworths. You got a china plate and cup and sat down, although the concept of eating standing up became popular at the Horn and Hardat Automat in New York; if you were in a hurry, "the company provided stand-up counters similar to those that banks provide for writing deposit slips. These people ate what became known as 'perpendicular meals.'"[13]

Then along came Dwight Eisenhower and the National Interstate and Defense Highway System, de-densification, sprawl, suburbia, and the two-car family. Colonel Sanders saw change coming, and when the Interstate 75 bypassed his restaurant, he sold it and started selling franchises instead. He introduced the bucket meal in 1957. Ray Kroc saw this coming and bought McDonalds. It was the start of the fast-food era, where you didn't get a china cup or a plate or even a seat. It was much more cost-effective to outsource the real estate to the customer's car, and to never have to buy or wash a plate again.

Then the constant upgrading began, when paper wrappers changed to Styrofoam clamshells. Because the cost of food was less than the cost of real estate and you didn't care anymore if people dawdled, portion sizes and cup sizes exploded. They trained us so well; we forgot how to actually sit for a 5-ounce cup of coffee, and wander around carrying a 20-ounce Vente. Our cars have multiple cup holders for every passenger and fold-down trays.

Suddenly we were in a linear single-use world, and almost everybody was happy; the fast-food operators made fortunes, the customers got convenience, the petrochemical industry sold more fuel to people driving and more plastic to wrap it all in. It became the North American way of life.

In his 1961 farewell address, Dwight Eisenhower famously described the *military-industrial complex*: "This conjunction of an immense military establishment and a large arms industry is new in the American experience.... Yet we must not fail to comprehend its grave implications."[14] In a less well-known portion of the speech, he worried that the nation was "giddy with prosperity, infatuated with youth and glamour, and aiming increasingly for the easy life":

> As we peer into society's future, we—you and I, and our government—must avoid the impulse to live only for today, plundering for our own ease and convenience the precious resources of tomorrow. We cannot mortgage the material assets of our grandchildren without risking the loss also of their political and spiritual heritage.[15]

He might have called this the *convenience industrial complex*. As Emelyn Rude noted: "By the 1960s, private automobiles had taken over American roads and fast-food joints catering almost exclusively in food to-go became the fastest growing facet of the restaurant industry."[16] Now we were all eating out of paper, using foam or paper cups, straws, forks; everything was disposable. And we still are. If anyone tries to change this (as when cities try and ban disposables), the next level of government will ban the bans because the petrochemical companies want to keep selling more plastic.

The Pandemic and Plastic

Prior to the coronavirus, there was some progress in banning plastic bags and you could get a china cup in Starbucks or Tim Hortons or bring your own, but that all went out the window at the first opportunity. John Hocevar, ocean campaign director at Greenpeace, told *CNBC*:

> The plastic industry seized on the pandemic as an opportunity to try to convince people that single-use plastic is necessary to keep us safe, and that reusables are dirty and dangerous. The fact that neither of these things is supported by the best available science was irrelevant.[17]

The pandemic also led to a dramatic increase in the amount of plastic used for take-out food. This was already becoming an issue with the rise of delivery services like Grubhub, Uber Eats, and Deliveroo, but their business boomed with the pandemic. And as one consultant noted, customers are finding that it is really convenient, which drives everything:

> "Yes, delivery is expensive...and if we do head into an economic downturn and money is a little bit tighter, I do think that will impact it," said Eli Portnoy, CEO of consumer research firm Sense360. "But the flipside of that is that over the last several months, consumers have gotten really used to the idea of convenience...and so I think we're still gonna see consumers use it at a much higher level than they were before."[18]

So, the big winners in the COVID-19 pandemic are the companies behind the convenience industrial complex, packing it all in plastic.

The Bottling Industry Goes Disposable

It wasn't just the fast-food industry that reaped the benefits of the Interstate Highway System. The Coca-Cola Company immediately saw an opportunity. Their business had historically been one of selling syrup to local bottlers who mixed it with soda water and sold it in returnable bottles, a perfectly circular system that only worked over short distances. Now, with big fast highways serving big suburban grocery stores where people could fill their big trunks with food, Coca-Cola could centralize production, capture all the money by supplying bottles, soda water, and syrup. They worked with Libby-Owens Ford to develop disposable bottles and with can manufacturers to develop cans that didn't make the acidic coke taste funny. Because they didn't have to worry about taking back the packaging, they saved a fortune; this was now the responsibility of the customer.

Since the customers didn't actually know what to do with it all, and were just throwing cans and bottles out the car window, the bottlers and the bottle manufacturers started the "Don't be a Litterbug" and "Keep America Beautiful" campaigns to teach consumers how to throw stuff away properly and, later, to recycle. Now the cities and towns that picked up the garbage were responsible, not the companies making the product.

Have a Beer

Changes in the brewing industry were even more disruptive. Beer used to be hyper-local, brewed and bottled in the communities where it was drunk, with the bottles being returned, washed, and refilled. But bottles were heavy, and two-way trips are expensive. When the Interstate highways opened, it became cheaper to build bigger centralized breweries, but canned beer in steel cans tasted terrible, and disposable bottles were expensive.

In 1959, Bill Coors invented the two-piece aluminum beer can, and instead of patenting it, offered it free and open-source to all the brewers. It didn't take long for Budweiser and Coors and Miller to build giant hyper-efficient centralized breweries and ship canned beer all over the country, to do massive advertising and price-cutting and put almost all of the local brewers out of business. Now, less than 3% of American beer is sold in refillable bottles. Every year on January 24, the fake holiday called Beer Can Appreciation Day, I write:

> The dominance of the American beer can is a story of the victory of centralized mass production instead of local, big business destroying small, the change from a reusable container to a disposable one, the switch from short-range shipping of a locally consumed product to the logistics of nationwide diesel transport, from healthy glass to BPA epoxy-lined aluminum cans. It is nothing to celebrate.

But it is all part of that convenience industrial complex, which almost forces us into a totally linear economy based on "take, make, and waste." It's structural, and it's cultural, and it is incredibly hard to change.

It's also getting worse. In Ontario, Canada, where I live, we had one of the world's best bottle recovery systems, with 85% of the beer sold in refillable bottles and 96% of those bottles returned. But the government thinks people want beer in corner stores, which will screw up the distribution and recovery system. Craft breweries are also booming, and they love cans; they don't have to worry about the bottle returns and get to put bigger labels on them. I yell at my own kids about buying beer in cans, but they just shrug; they like the microbrews.

When the pandemic hit, people stopped returning bottles, and the brewers ran out of them; now when you go into a beer store, all you see are cans. The kids say, "So what? Aluminum cans can be recycled." But only 65% of them actually are. Since there isn't enough recycled aluminum, demand for cans creates demand for new virgin aluminum, with its huge carbon footprint. And even recycling cans has its own costs; As Carl Zimring wrote in his book *Aluminum Upcycled*:

> Although the contaminants released by recycling pale compared to the ecological damage of mining and smelting primary aluminum, the waste products of scrap recycling must be considered when considering the consequences of returning the metal to production.[19]

Water, the Ultimate Scam

In many cities (certainly not all), municipal water systems are irreplaceable treasures. New York gets its water from a vast 2,000-square-mile protected watershed, some from the 1890s and some assembled through eminent domain, eviction, and displacement in the 1940s. Vancouver and Seattle drink fresh mountain water collected in reservoirs, checked many times a day. Toronto's water is treated in a cathedral to water, one of the most beautiful buildings in the city; it is tested over 20,000 times a year, and the results are public.[20]

So why, in every one of these cities, do you see people walking around drinking bottled water? Elizabeth Royte, in her book *Bottlemania*, called it "one of the greatest marketing coups of the 20th century,"[21] and it continues well into the twenty-first. She described how a PepsiCo marketing VP told investors back in 2000: "when we are done, tap water will be relegated to showers and washing dishes."[22] Or as the Attorney General on Idiocracy wondered about drinking water instead of Brawno, "Water? Like out the toilet?" She also explains how the industry will deal with the question of the plastic bottles, quoting a Coke exec: "Our vision is to no longer have our packaging viewed as waste but as a resource for future use."[23]

But people don't need all that much water, so they had to be convinced that you need to stay hydrated and drink eight glasses a day, which simply isn't true.

> "The control of hydration is some of most sophisticated things we've developed in evolution, ever since ancestors crawled out of sea onto land. We have a huge number of sophisticated techniques we use to maintain adequate hydration," says Irwin Rosenburg, senior scientist at the Neuroscience and Ageing Laboratory at Tufts University in Massachusetts.[24]

In his recent book, *Less Is More*, Jason Hickel blames the capitalist system for the takeover of a public resource.

> For instance, if you enclose an abundant resource like water and establish a monopoly over it, you can charge people to access it and therefore increase your private riches. This would also increase what Maitland called the "sum-total of individual riches"—what today we might call GDP. But this can be accomplished only by curtailing people's access to what was once abundant and free. Private riches go up, but public wealth goes down.[25]

Of course, all those bottles are a solid fossil fuel. A 2015 study[26] found that the production of plastics contributed 1.8 gigatonnes of CO_2e, or about 3.8% of global emissions, and 42% is for the convenience of single-use packaging, much of it making the 583 billion bottles that are estimated to be sold in 2021.[27]

Meanwhile, the petrochemical industry wants us to buy more plastic. They are concerned about electric cars reducing the need for gasoline and are pivoting to plastic, investing $400 billion in new production facilities in the next five years. They are doing everything they can to ensure that we keep buying what they are making. Look forward to the coronavirus being used as an excuse to continue not having reusable or refillable cups at Starbucks and Tim Hortons, or refillable bags at the grocery store. We will be taught that anything that isn't wrapped in plastic might be tainted. But so much is at stake; according to Carbon Tracker:

> Carbon dioxide is produced at every stage of the plastic value chain—including being burnt, buried or recycled, not just extraction of oil and manufacturing. The analysis therefore finds that plastic releases roughly twice as much CO_2 as producing a tonne of oil.[28]

The Carbon Tracker report suggests that "there are three main solutions—reduce demand through better design and regulation; substitute with other products such as paper; and massively increase recycling."[29]

Goodbye Recycling, Hello, the Circular Economy and Chemical Recycling

Recycling was the industry's answer to the landfill crisis of the seventies, the idea that it would all be picked up and made into something else. But, as we learned, it was all a sham as it was all shipped to China where the labor was cheap enough that they could afford to separate it, and the environmental regulations were lax enough that they could burn or landfill what they couldn't find any value in. It was a problem when China closed its doors, but a bigger problem was the fracking revolution and the vast amount of natural gas that came onto the market with no place to go. So, while the cost of collection and processing of plastics like PET (the stuff of water bottles) has remained the same, according to *Recycling Today*, "virgin PET prices have been driven down by a series of developments: competition boosted by domestic overcapacity, flattening of demand, pressure from lower cost imports and plummeting raw material prices."[30]

The plastics industry is adapting to the changed conditions by co-opting "the circular economy," claiming that they will use "chemical recycling" to process plastics back to their original constituent chemicals so that they can be made into what are essentially virgin plastics. Except they still have the problem of collection, and chemical recycling has a huge carbon footprint of its own. It's really just a way to promote waste-to-energy (incineration) and to tell the governments and consumers "Don't worry, we're doing something!"

Change the Culture, Not the Cup

The circular economy is all the rage these days. As described by the Ellen MacArthur Foundation, it involves *designing out waste and pollution*: "By changing our mindset to view waste as a design flaw and harnessing new materials and technologies, we can ensure water and pollution are not created in the first place."[31] And *keeping products and materials in use*, again, more elaborate recycling, and *regenerating natural systems*.

The problem is that it is really hard to bend a linear economy into a circle. It is designed around convenience and waste, and around the concept that the customer is buying the container, not the contents. With water, the bottle is the product, an expensive way of carrying tap water that the customer has already paid for with their taxes. The linear economy is so profitable because the companies have managed to get someone else to pick up the tab for the waste collection, the street cleaning, the disposal. They have convinced citizens that it is a virtue to dress up in yellow vests and spend their time collecting bottles and coffee cups from the side of highways.

You can try to be circular in a linear economy. You can hope that customers will return their empty cups to a store so that they can be transported to a reprocessing plant, pulped, separating the poly liner from the cardboard, turn that into more cardboard, and try and sell the plastic, but it really isn't worth the trouble.

The real solution is to change the culture, not the cup. Sit down in a coffee shop instead of getting takeout to drink on the street or in your car. If you are in a hurry, drink like an Italian: order an expresso and knock it back, standing up.

The linear economy was an industry construct that took 50 years to train us in this culture of convenience. It can be unlearned.

The Stuff That We Buy

Electronics

In 2019, 161 million smartphones were sold in the US,[32] each with an average embodied carbon footprint of about 66 kg of CO_2e. That's about 10.66 million tonnes of CO_2e, the equivalent of 1.6 million flights from New York to London. On average, Americans replace their phones every 24.7 months, which is less often than they used to. Phones are such cute little things, but we ignore how much energy goes into making them. Vaclav Smil did an interesting analysis where he compared phones to cars and found that "new cars weighed more than 180 times as much as all portable electronics but required only seven times as much energy to make."[33] Then he took their comparative lifespans into account and found:

> Portable electronics don't last long—on the average just two years—and so the world's annual production of these products embodies about 0.5 EJ [exajoule, or quintillion joules] per year of use. Because passenger cars typically last for at least a decade, the world's annual production embodies about 0.7 EJ per year of use—which is only 40 percent more![34]

And yet I complain bitterly about cars (and kept my last one for 16 years) and am on my fourth iPhone. I have had my priorities completely wrong. And it gets worse.

Back in 2012, my old computer was dying, and everyone I knew who wrote was using Apple Macs, which were a lot more expensive than my home-built PCs and cheap notebooks that never seemed to last more than a year or two. I called an architect and writer friend in Miami for advice, and he said, "Buy the best they have, get three years of Apple Care, and when it is up, buy a new one." His point was that this is your livelihood and you need something good with a good warranty.

In 2020, that 15" MacBook Pro was still working perfectly. The battery was down to 80%, and the latest Apple operating system upgrade was the first that it couldn't handle, but it could still do everything I needed, with a big crisp screen and great keyboard that hasn't been significantly improved upon.

But somehow that wasn't good enough for me. I wanted bigger and better, so I bought a desktop, a 27" iMac in 2016. I also wanted smaller and lighter for when I was traveling, so I bought a super-light MacBook. I am reading constantly, and an iPad was way more convenient. And of course, I had to have the new iPhone 11 Pro because the camera is so great and I have been waiting forever for a wide-angle lens, a necessity for architecture.

So here I am today, owning a catalog of just about everything Apple makes, and that original MacBook Pro is still cranking away; I have lent it to a friend who is running a pizza shop with it. But had

I not been so enamored with the latest and shiniest, I could still be working on that same machine, on a better keyboard with a bigger screen than the MacBook Air I am writing this on.

The problem is the embodied carbon in making all this. Apple recently released its latest environmental report that includes the carbon footprint of all of its products,[35] and all my Apple stuff adds up. In Apple's analysis, my iPhone 11 Pro has a lifecycle footprint of 80 kg of CO_2e, with 83% coming from production, the embodied carbon. This is not just talking about what happens in the factory but in fact "includes the extraction, production, and transportation of raw materials, as well as the manufacture, transport, and assembly of all parts and product packaging." Use is based on a "three-year use by a three-year period for power use by first owners. Product use scenarios are based on historical customer use data for similar products" and is 13%. The footprint is based on an average power grid mix, so since my electricity is carbon-free, I can probably discount this, bringing the phone's footprint down to 69 kg, which divides out to 63 grams of CO_2e per day. I can live with that.

But it's clear that embodied carbon just dominates the footprint of portable electronics, and I have a lot of them. I am writing this on a MacBook Air (I bought it when I travelled a lot! I needed a really portable machine!)—174 kg. There is my Apple Watch (fitness tracking! Gotta have it!)—40 kg for the tiny little thing. My iPad (I read a lot! It's great for consuming information!)—119 kg. Then there is the elephant in my room: that 27" iMac, with a carbon footprint of 938 kg, almost a tonne. They are all extraordinarily efficient devices, and if you are just measuring your operating emissions, they can almost be ignored. But it takes so much stuff and energy to make them, that 80% of their footprint is upfront in the embodied carbon (except for the iMac, which is only 45% upfront; it takes more power to light up that big screen).

And as I noted, almost none of this was necessary, I just kept getting seduced by the shiny and new. The bigger and then the smaller. I could still be perfectly happy writing on the MacBook Pro with an external monitor; I did it for years.

It's just more evidence of my earlier point that thinking about embodied carbon changes everything. In a car or a house, operating emissions dominate; in electronics, they barely matter, it's all in the embodied carbon.

The footprint of just my Apple paraphernalia totals 1.271 tonnes, and even spread over three years, this is a fifth of my carbon budget. On a daily basis, it's 0.891 kg, almost as much as gas and electricity, and it is only that low because I am pushing the iMac into five years. It is a significant part of my budget, about twice what it would be if I had been more sensible and disciplined. And I am going to push my Apple collection as far as it can go, because every year I get out of them reduces the impact of their embodied carbon. But here's the lessons I learned:

- Buy good-quality stuff, treat it with care and make it last.
- There was a time where every three years computers were obsolete. It's not true anymore, the changes are incremental. Keep it going as long as you can.
- My new iPhone 11 Pro is really the first phone that improves on my first iPhone, the elegant little 4S—I really didn't need anything in between.

Clothing

About 30 years ago, my wife's grandmother was moving out of her apartment and into a nursing retirement home and pulled out a suitcase for me. Her husband, Hugh, who had died about 30 years before, had a gambling problem and every Friday would buy a new tie for good luck at the track. When he lost (which he usually did), he put the tie away and never wore it again. So, I was suddenly the proud owner of a fabulous collection of dozens of skinny ties from the 1950s that I wore constantly until people stopped wearing ties. I still have them all for when ties come back into fashion again.

Being an architect, it is all very easy; just buy a lot of black stuff and you are fine. My black suits are all at least ten years old, and I haven't worn one in at least five years. I didn't consider clothing to be a big part of my carbon footprint. But for many people, it can be significant. I was shocked to learn the scale of the clothing and textile industry.

The C40 Cities report on the future of urban consumption claims that our clothing and textiles were responsible for 4% of our urban footprint surprised me.[36] But in fact, it's much greater than that. Worldwide, the size of the clothing industry is vast, employing 300 million people. A single pair of jeans has a carbon footprint of 33.4 kg of CO_2e; the entire industry pumps out 3.3 billion tonnes of CO_2e, or about 8% of total global emissions. That's bigger than either steel or concrete. Or as Greenpeace notes, "flying has nothing on the carbon footprint of the fashion industry."[37]

Of the fast fashion sold today, 60% is made from polyester, which we have described as a solid fossil fuel; a single polyester shirt has a carbon footprint of 5.5 kg CO_2e. Much of this ends up in the ocean; according to a World Bank article:

> Every year a half a million tons of plastic microfibers are dumped into the ocean, the equivalent of 50 billion plastic bottles. The danger? Microfibers cannot be extracted from the water and they can spread throughout the food chain.[38]

The article notes also that it doesn't last very long:

> The average person today buys 60% more clothing than in 2000, the data show. And not only do they buy more, they also discard more as a result.... The Ellen MacArthur Foundation estimates that every year some USD 500 billion in value is lost due to clothing that is barely worn, not donated, recycled, or ends up in a landfill.[39]

Another study found that clothing was designed to last "not much longer than ten wearings."[40]

The fast-fashion business was transformed by the COVID-19 pandemic; sales dropped so quickly that it caused a humanitarian crisis in Bangladesh with 1.2 million workers affected.[41] Many in the industry believe it will take years to recover because of store closures, reduced incomes, more people working from home, and changes in taste, with everyone living in their athleisurewear.

Perhaps it is a good time to rethink what we wear. Kelly Drennan, Executive Director of Fashion Takes Action, suggests a bunch of Rs that go beyond the usual three:

- Reduce: "shop for value instead of cost."

- Reuse: buy secondhand and vintage.
- Repair: everyone used to do this, from sewing buttons to turning collars.
- She also suggests researching which companies have the best track records, renting, and repurposing—"T-shirts into totes!"[42]

We could also emulate the late Steve Jobs, who famously had a pile of identical mock turtlenecks and jeans and just wore the same thing every day, or just dress like an architect.

Furniture

I was shocked when I picked up my daughter from her student apartment near the University of Guelph at the end of term; at the curb in front of every building there were mountains of junked IKEA particle board furniture. Desks and chairs were essentially now disposable, use one season and throw it away. IKEA changed our attitudes toward furniture; it used to be that young people started with hand-me-downs from Grandma or used furniture. The stuff was heavy and hard to move, but it was also expensive when new. Now it is cheaper to buy an IKEA sofa than to rent a truck or hire movers to get a sofa from Grandma's house. The modern Scandinavian design that used to be found only in high-end shops is now the universal standard. IKEA took over the world. After years of criticism for its forestry practices, IKEA now only uses wood that is certified as sustainably harvested, and as we noted earlier, wood is a renewable resource when managed properly. They also do sell some more durable products.

But we have become slaves to the IKEA style. Architectural critic Edwin Heathcote was talking about Airbnbs, but this description applies to almost every interior these days:

> They are designed to seduce with an idea of generic global familiarity. They represent a lifestyle that is metropolitan, chic, minimal and self-congratulatory. You enjoy the image because this is how you imagine you might want to live. You recognise it already. What is happening here is a kind of invidious digital

> aesthetic seepage, an unintentional effect of the gradual global convergence of the interior.[43]

The old-fashioned "brown" furniture that our grandparents loved all goes to the dump now; people can't give it away, even though it is so solid it can last for generations. It just doesn't fit that modern aesthetic. (It's also often too big for today's little apartments. That's why IKEA dressers kill children[44]—they are not deep enough to be stable, but they are shallow enough to fit in tiny bedrooms.) I was lucky that my late mother was a designer and was buying mid-century modern furniture back in the 1950s, so I have been happy with hand-me-downs.

In fact, perhaps the only category in this exercise where I have always done the right thing is in furnishing; I have almost always bought vintage furniture and keep it all forever. I am writing this while sitting at a table made from an old bowling alley that I renovated as an architect 30 years ago; the wood was probably 30 years old at the time. It took me 30 years to find dining room chairs that I liked; they were vintage fifties designs.

There are many wonderful things about vintage furniture. It is often cheaper than new, though really good stuff can be expensive. It is usually solid wood, with no formaldehyde or outgassing. It's built to last, although all my Charles Eames chairs have fallen apart as the rubber pucks that connect to the legs dried out. It can even be a good investment; when we downsized; I sold a classic Herman Miller desk and stool for a huge profit, after using it for 30 years.

What Do We Learn from the Spreadsheet?

This was a difficult year to get a real picture of consumer goods. I didn't buy any clothing or much at all, but I am seriously paying the price in embodied carbon for all of my Apple stuff. I am going to make what I have last as long as Apple keeps supporting it. It's also time to think about my daily newspaper; I subscribe to a paper version because that's how you help keep the newspaper business alive, but since the pandemic, there is almost no advertising, it is a shadow of its former self, and I might as well just read the digital version at this point.

Table 8.1. Calculations for consumer goods

Item		Source	Unit	CO_2/ unit	Estimates/ day	Nov 2	Nov 9	Nov 25	Nov 26	Nov 27	Nov 28
Newspapers	weekend	berners-lee		1800	900	400	400	400	900	1200	400
Plastic			grams CO_2	6							
Apple embodied					891.14	891	891	891	891	891	891
Leisure											
TV Time			grams/hr		50.4		25	100.8	101		
Services											
Data	variable	JC Mortreux	gms/GB	123	62						
Data	revised	finish study and ot	gms/GB	10	5	60	75	40	40	40	75

What Can We Do?

From computers to clothing, the question about sufficiency applies: how much do we really need? It appears that, for any consumer good, the best strategy is to buy high quality with timeless design, maintain it well, and use it for as long as you can.

Leisure

From the 1.5-degree lifestyle report: "leisure activities performed outside of the home, e.g., sports, culture, entertainment, hotel services." Emissions from ingredients of food taken out of home are included in nutrition, whereas direct emissions from leisure performed at home are included in housing.

Why Do We Have Leisure?

According to the 1.5-degree lifestyle study, leisure isn't one of the three "hotspots." When I started with this category, I didn't think it would be very significant either. In fact, the more one looks at leisure, the more important it becomes, especially in a world where we are considering the question of sufficiency, of how much we really need, and much we actually work.

In 1899, Thorstein Veblen could title his book *The Theory of the Leisure Class* because there was, in fact, a class of people who had time on their hands, as distinct from those who had to work. There was not only conspicuous consumption but also conspicuous leisure, which he defines as the non-productive consumption of time. Members of the working classes didn't have much time for leisure, but the rich wanted to demonstrate how much leisure they actually had, thanks to all the people working for them. "Conspicuous abstention from labour therefore becomes the conventional mark of superior pecuniary achievement and the conventional index of reputability."[45] The leisure class couldn't just have it, they also had to flaunt it: "The wealth or power must be put in evidence, for esteem is awarded only on evidence." This may well explain the consumption of Sea-Doos and power boats and ATVs, all those expensive and powerful toys that people use to fill their leisure time.

I immediately thought of a personal example, from my days working in real estate development. All the other developers belonged to ski clubs, so of course I had to join one, first skiing (I was terrible) and then snowboarding. Every weekend in winter, I would pack the kids into the Outback and drive two hours north to the ski hills in Collingwood, Ontario, where we would get electrically cranked up to the top of the escarpment for a quick slide on the artificial snow made the night before with huge amounts of electricity and water pumped from Georgian Bay. As the climate warmed, the seasons became shorter and shorter, but nobody up there in their big SUVs and fancy chalets connects their actions to this.

Looking back at it now, it was pure Veblen, conspicuous and very expensive leisure. When that career ended inauspiciously, I gave up the club, and came to understand the silliness of spending four hours in a car to have four hours going up and down hills, spewing carbon all the way; now I cross-country ski in a city park. That's a serious *modal shift*, and not nearly so conspicuous.

Since Veblen was writing back in 1899, there have been radical changes. With the reduction in hours of work and increases in pay after the First World War, many more people had time on their hands. Economists thought it would continue this way; in 1928, John Maynard Keynes predicted that because of a predicted eightfold increase in productivity, we would be working on average three hours a day by 2030. He thought it would make us better people:

> We shall once more value ends above means and prefer the good to the useful. We shall honour those who can teach us

> how to pluck the hour and the day virtuously and well, the delightful people who are capable of taking direct enjoyment in things, the lilies of the field who toil not, neither do they spin.[46]

Bertrand Russell wrote in his 1932 essay "In Praise of Idleness" how everyone should now be a member of the leisure class:

> Leisure is essential to civilization, and in former times leisure for the few was only rendered possible by the labors of the many. But their labors were valuable, not because work is good, but because leisure is good. And with modern technique it would be possible to distribute leisure justly without injury to civilization.[47]

He thought that everyone should work fewer hours and have more time for leisure. The problem that he foresaw and that we see today is the issue of "positive idleness" such as painting or writing vs "negative idleness" which "ends up being the effect of work under the spell of consumerism and its consequent socioeconomic inequality."[48]

> The pleasures of urban populations have become mainly passive: seeing cinemas, watching football matches, listening to the radio, and so on. This results from the fact that their active energies are fully taken up with work; if they had more leisure, they would again enjoy pleasures in which they took an active part.[49]

We are all so worn out from our day at work that we basically flop down in front of the television. This is borne out by the statistics, which is why the artificial division of this category is distracting. The 1.5-degree lifestyle study includes "leisure activities performed outside of the home, e.g., sports, culture, entertainment, hotel services," but in fact we spend the vast majority of our leisure time at home.

And why didn't we get all that extra time that Keynes and Russell promised us? According to Benjamin Freidman, it's because of inequality. Basically, all the gains from the increases in productivity went into the pockets of the rich. He quotes economist James Meade, writing in 1965, discussing how having more time would have the opposite effect:

> Meade instead thought "wages would thus be depressed," as ever less labor was necessary for production. Correspondingly, an ever greater share of total income would go to the owners of the machines. In the absence of government-provided welfare on a massive scale, therefore, most of the workforce would be compelled to take whatever low-paying jobs they could get, presumably in the service of the machine-owners but not working with the machines. In Meade's vision, "we would be back in a super-world of an immiserized [impoverished] proletariat of butlers, footmen, kitchen maids, and other hangers-on." In today's American context a half-century later, one might substitute gardeners, swimming pool attendants, personal trainers and home nurses.[50]

Which brings us back to the concept of degrowth. As Jason Hickel wrote in *Less Is More*, it's all about distributing income and resources more fairly, liberating people from needless work, and "investing in the public goods that people need to thrive."[51] He continues, sounding a bit like Bertrand Russell:

> Some critics worry that if you give people more time off they'll spend it on energy-intensive leisure activities, like taking long-haul flights for holidays. But the evidence shows exactly the opposite. It is those with less leisure time who tend to consume more intensively: they rely on high-speed travel, meal deliveries, impulsive purchases, retail therapy, and so on. A study of French households found that longer working hours are directly associated with higher consumption of environmentally intensive goods, even when correcting for income. By contrast, when people are given time off they tend to gravitate towards lower-impact activities: exercise, volunteering, learning, and socialising with friends and family.[52]

This is why leisure will become an increasingly important part of our carbon picture. In societies with strong social safety nets and less inequality, people have more leisure time and less worry. According to the International Labour Organization, "Americans work 137 more hours per year than Japanese workers, 260 more hours per year than British workers, and 499 more hours per year than French workers."[53]

The ideas of degrowth, sufficiency, and more leisure time all begin to look very attractive.

What Exactly Is Leisure Today?

The American Time Use Survey[54] prepared by the U.S. Bureau of Labor Statistics found that we have a lot of time for leisure, averaging 5.19 hours per day. Over half of that, 2.81 hours, is spent in front of the television. Add in time in front of the computer surfing or gaming (0.55 hours for men, 0.31 for women), and you have a lot of time looking at screens. Bertrand Russell would be unhappy to know that we only spend 0.32 hours reading or thinking, 0.64 hours socializing and communicating, and 0.31 hours on sports, exercise, and recreation, with a real split between men (0.39 hours) and women (0.23).

When I started this exercise, I expected that I would be writing long diatribes about people who drive all-terrain vehicles and snowmobiles and big motorboats instead of bikes, skis, and canoes. But when you study the data, it becomes clear that this is a very small proportion of the population who have the conspicuous leisure and the money to buy toys that they actually don't use that often.

In fact, most of our leisure time is spent sitting on a couch looking at a TV or some other screen. If we apply the same rules of consumption vs production that we do for our Haier air conditioner, then we have to include our share of the carbon footprint for the entire entertainment industry. What's behind the screen during those 2.81 hours of TV? Lauren Harper of the Earth Institute wrote:

> The United States film and entertainment industry produces an average of 700 films and 500 television series a year. On average, these industries spend millions of dollars on everything from flights for actors and actresses, to food for crew teams, fuel for trailer generators and, of course, electricity for picture perfect light. While this results in award-winning entertainment and enjoyable evenings of episode binging, these productions can have large carbon footprints and significant

> environmental impacts. For example, movies with a budget of $50 million dollars—including such flicks as *Zoolander 2*, *Robin Hood: Prince of Thieves*, and *Ted*—typically produce the equivalent of around 4,000 metric tons of CO_2.[55]

The entire entertainment industry is moving into our TV room, with Netflix, Apple, and Amazon Prime producing thousands of hours of entertainment that comes straight into our homes, and one could probably write another book just about its footprint. Netflix et al. are producing it for our consumption, but I don't know where to start in measuring it. From *Vice*:

> According to BAFTA, the British film organization, a single hour of television produced in the U.K.—fiction or nonfiction—produces 13 metric tons of carbon dioxide. That's nearly as much CO_2 as an average American generates in a year. A 2006 UCLA study found that the California film and television industry created 8.4 million metric tons of carbon dioxide; the number for the U.S. film and TV industry as a whole was 15 million tons.[56]

If Bertrand Russell thought we had a problem of passivity in 1932, he would be truly shocked today. There is a giant pile of carbon behind every minute of our screen time, which is why it is increasingly important that we turn it off. The real opportunity in reducing one's footprint of leisure is to substitute as many minutes of passive leisure for active, to get off that couch and get outside. This adds years of healthy living and puts you in touch with your community, making it stronger.

And for all the people who say that our personal actions don't matter but our collective actions do, perhaps some of that TV time could be put to better use volunteering and taking to the streets.

What Can We Do?

- Pick active non-motorized activities, cross-country over downhill skiing; canoe instead of motorboat; bike over ATV.
- Walk to your local neighbourhood restaurant or bar instead of driving to the hotspot across town.

- Don't trade up to the latest bigger TV, most of the footprint is embodied; keep your old one going as long as possible. Better still, turn it off.
- Get off Twitter and read a book. Doing this while researching this book has actually changed my habits and got me through those 1,000-page Vaclav Smil doorstops.
- At every opportunity, try and spend your leisure time moving, doing something active, instead of sitting in front of a screen. It may not make that big a difference to your carbon footprint in the short term, but it will make a big difference in your health.

Services

From the 1.5-degree lifestyle report: "services for personal purposes, e.g., insurance, communication and information, ceremonies, cleaning and public baths, public services." Public services covered by government expenditure are excluded from lifestyle carbon footprints.

How We Connect

When I started this exercise, I thought that one of the biggest numbers in my carbon footprint was my communication, my internet use, which I calculated by multiplying the daily gigabytes my internet provider reported by 123 grams per GB, a number I got from various sources, which were all based on data from about 2010. In the last decade, we have been deluged with new internet services, streaming companies, and now, of course, Zoom meetings. Many people (like me) have offloaded their storage to the cloud, both for security and convenience. The number of gigabytes everyone is burning through has soared exponentially. One might think that my estimate might actually be low.

But energy is a major operating cost, so the companies have been ruthless in their hunt for efficiencies. The servers and the hardware have followed a Moore's Law-like increase in efficiency and reduction in energy consumption per gigabyte handled. It really had to, or Google and Amazon would be sucking up every kilowatt in the country. Cooling the data centers was one of the biggest consumers of electricity, so they located many of them in cooler places and switched to chips that put out much less heat.

Meanwhile, the data companies got greener. Apple claims to run the iCloud on 100% renewables, Google claims to be carbon-neutral, as does Microsoft. Netflix "offsets and buys renewable energy certificates." Amazon, by far the largest cloud service, promised to be 100% renewable but is really only about 50% now and has been backsliding.

While we have a choice of which streaming service to watch, it really doesn't make much difference; most of the internet seems to run on Amazon Web Services these days. But even if they are a bit dirtier, the fact is that the overall mix of how data centers are run and the equipment they use means that the carbon footprint per gigabyte that I am using in my calculations has dropped from 123 grams per GB to somewhere between 6 g and 20 g. Since I live in Ontario and my internet provider, Bell, is using Ontario and Quebec low-carbon power, I am going to go on the low side with 10 g. This has made a huge difference in my daily footprint.

Meanwhile, poor Rosalind Readhead has been counting every email, text message, and phone call in her daily ledger, and they are probably off by a factor of ten as well, they are barely rounding errors now. She shouldn't bother.

There is the power used to run our phones too, which is also negligible; they are very efficient devices. This is not out of any desire to conserve energy or reduce carbon; it is a design necessity, to have a small phone that lasts all day. Apple estimates that a modern phone uses about 70 cents worth of electricity per year. The carbon footprint of operating a phone is barely relevant. What matters is the footprint of making it.

Education

Ever since I started teaching at Ryerson University ten years ago, I have complained about what an archaic system it was. When the first lecturer stood up at the front of a class at the University of Bologna in 1088 and started reading, it was because books were written, not printed, and students didn't have them. So you learned by listening to someone read from a big book. And here we are, 930 years later, and we still have what is often some old white guy standing up at the front of the class talking away with students making notes. I thought surely, in this modern age with all our technology, we could do better than this.

Others did too; a decade ago every university was chasing MOOCs, or Massive Open Online Courses; 2012 was called "The Year of the MOOC." There were legitimate concerns that, as David Glance of the University of Western Australia wrote, "It was really only a matter of time before bricks-and-mortar universities faced the same digital battle-to-the-death that has transformed other industries like newspapers and music. Universities faced a MOOC 'tsunami' and 'avalanche.'"[57]

It didn't happen because it turned out that students are not interested in viewing lectures, they are interested in getting degrees, and the universities were not interested in giving them out on the basis of online courses when they had so many bricks-and-mortar buildings to fill up and they wanted that full tuition. As Glance notes, there was no business model that worked:

> Universities are not free enterprises in the same way as music companies and news media companies are privately owned companies. Universities are governed by legislation and quality requirements that restrict who can award a degree and how they have to go about ensuring that students achieve a particular standard in order to earn that degree. This has made any change to this model a matter for governments as well as universities and their customers.[58]

How quickly that changed with the pandemic! Suddenly, almost overnight, online courses met those quality requirements and standards because universities were desperate to get the tuition and the students are desperate to get those degrees.

Running a university has a massive carbon footprint not only in its real estate but also in transportation for the students. Many in my class would travel for two hours on bus and subway each way, just for my two-hour class, where they rarely asked questions and often seemed to be just looking at their phones. If you are just doing a straight lecture, a professor talking and a student watching, who needs this?

The pandemic is going to cause the breakthrough in online teaching that the MOOCs couldn't because the good teachers have learned how to deliver their lessons, how to deal with students online, how to adapt to a challenging situation. When the pandemic is over and everyone is vaccinated, it will not go back to the way it was before but will probably be a hybrid model where students can do either; the time when we pack 350 students into a lecture hall is probably gone forever. And, like the move to working at home, it will eliminate vast amounts of carbon emissions.

The Coffeeshopification of Everything

In 1895, the Canadian humourist Stephen Leacock described going into a bank:

> When I go into a bank I get rattled. The clerks rattle me; the wickets rattle me; the sight of the money rattles me; everything rattles me. The moment I cross the threshold of a bank and attempt to transact business there, I become an irresponsible idiot.[59]

We all used to go to banks, line up during their limited hours, and get a bit intimidated. And today? We don't actually go to banks to do transactions, we do almost everything online. When we do go in, there are comfy chairs, maybe coffee, and it looks a lot like—a coffee shop.

Almost a decade ago, the futurist Steven Gordon was watching all these trends of MOOCs and telecommuting and predicted that "in the future, everything will be a coffee shop." In universities, "you could have local campuses becoming places where students seek tutoring, network, and socialize—reclaiming some of the college experience they'd otherwise have lost."[60]

Offices are turning into coffee shops as they reopen after the pandemic; people can work from home if they like, but they come in for meetings and important discussions. The major function of the office is to interact, to schmooze. Like you do in a coffee shop.

Stores are also being coffeeshopified; they want customers to come in and stay awhile, develop a bond, feel comfortable and at home. The few bookstores that are left became coffee shops years ago; other brick-and-mortar stores are following.

This is the low-carbon future: we do everything on our phones or our computers, minimizing transportation and real estate. When we need to interact with people in real life, whether it is an office or a school or a business, it will feel like a coffee shop.

What Can We Do?

- Courtesy of the pandemic, you can do almost every service online these days. I used to visit my accountant at year-end; now I send her PDFs and scans. Many are finding this to be convenient and effective, and every service that is done this way saves carbon.

–9–

Conclusion: In Pursuit of Sufficiency

Shortly after I started writing this book, everything shut down because of the COVID-19 pandemic, and my spreadsheet tracking my carbon footprint became very boring. I really didn't have much of a problem living a 1.5-degree lifestyle because I didn't have much of a lifestyle at all; every day was pretty much the same. The times I broke 6.8 kilograms of CO_2 per day were obvious: the single red meat dinner, two trips up to my cabin and back, and one trip into town from the cabin to get my keyboard fixed. My spreadsheet was monotonous, especially all summer, when most of the traveling I did was a walk to the end of the dock.

I did more than my usual lecturing, but didn't get to New York, Vancouver, Washington, Berlin, or Lisbon in person; it was all by Zoom. I suspect that many people were in the same situations, particularly those lucky enough to have jobs where they can work from home and still earn their full salary. It's relatively easy to live a 1.5-degree lifestyle if you can't fly and have nowhere to drive to.

Others did not have such a low-carbon pandemic. Many sat for hours in their idling cars that have become PPE (personal protective equipment), waiting to be tested for the virus or to get help at food banks. Many bought first or second cars so that they could drive to work because transit had become so spotty and suspect. Used car sales boomed 22% in 2020; one nurse told the *New York Times* that her family bought a second car so she could avoid transit: "We used to take Uber to restaurants, especially if we were going out with friends and didn't want to drink and drive," said Ms. Cray, who has treated patients with the coronavirus. "We don't do that anymore. We take our car instead."[1]

The newspapers were full of stories about how people are buying houses in the suburbs and the country, wanting more space inside and out. Most planned to mainly work from home and drive into the city for meetings or doctors' appointments. They were buying bigger houses that can do everything; real estate agents in Florida said people were looking for "home offices, larger kitchens for increased in-home dining, home gyms and private pools."[2]

Home delivery of food skyrocketed as restaurants closed, fueled also by Silicon Valley-funded apps, and people liked it. A consultant noted that "over the last several months, consumers have gotten really used to the idea of convenience...and so I think we're still gonna see consumers use it at a much higher level than they were before."[3] All of this is delivered in single-use plastic containers.

So, while the pandemic caused a dramatic decline in emissions for half of 2020, it could be that they are going to come back bigger than ever as the pandemic recedes, or alternatively, things may be very different.

It might well be like the Roaring Twenties 100 years ago, where pent-up demand after years of war and then the flu pandemic led to an explosion in consumption. That decade was crushed by the 30s and the Great Depression, and ours might be as well by our failure to deal with the climate crisis.

But for others, the world has changed because of the pandemic, jobs were lost, and they are rebuilding their lives. Living the 1.5-degree lifestyle can help do this; it is also a frugal lifestyle. Food is cheaper and can be healthier, transportation becomes exercise, consumption of unnecessary things is lower. It may well be that a low-carbon frugal lifestyle becomes a way of life for many.

When I started this book, I broke my footprint down into six neat categories because it seemed like a logical thing to do at the time, given the 1.5-degree lifestyle report that inspired the exercise used them. I had doubts about what the 1.5-degree lifestyle study called "hotspots"—nutrition, housing, and mobility, basically what we eat, where we live, and how we get around. In fact, it turned out to be completely accurate. The more I learned, and the more I refined the spreadsheet, the more sense the hotspots made. For instance, I started including the footprint of domestic water, recalculated the electricity based on more recent numbers, and pretty soon just the utilities—the gas, electricity, and water—were taking up a third of my daily carbon budget.

But it quickly became clear that life doesn't break down into six neat independent silos; they are all versions of the same thing, just speaking in different languages. Because of where I live, it is easy for me to avoid driving. But even the carbon footprint of what we eat changes dramatically because of where we live, whether our daily bread is a baguette baked in the boulangerie around the corner or one made from frozen dough cooked up at the Walmart. It's hard to change little things when everything is interconnected; it requires a different way of thinking.

Superficially, tracking my carbon told a simple story: Don't eat red meat, don't drive a gasoline powered car, don't get on an airplane. But when you dig in, it told a much bigger story, and there were many lessons that popped out of the spreadsheet:

- The embodied carbon, the upfront carbon emissions from making stuff, is as or more important than the operational emissions and becoming more so every day. My electronics use almost no energy to run, but the carbon emitted while making them has buried me. The same is true of our homes, cars, and buildings and just about everything that is made of metal or silicon or concrete.
- How we live and how we get around are not two separate issues; they are two sides of the same coin, the same thing in different languages. It's much easier to live a low-carbon life if you live in a place designed before the car took over, be it a small town or an older city. But for the people who don't do that, the problems are immense.
- We don't need a radical change in diet; we don't all have to go vegan. But it is not such a big deal to eat a lot less red meat, lamb, and dairy; to eat appropriately sized portions; to be more careful about waste; to eat local and seasonal and shorten that cold chain.
- The way we work has changed because of the pandemic, and many of those changes save a lot of carbon. The reductions in commuting may well be permanent; most people would rather Zoom than commute. The way we communicate has changed; I have been seeing a lot more people from more places on my

Zoom screen than I ever saw in person, and some meetings that used to happen annually are now happening every week. The Third Industrial Revolution is here, and if we do it right, it could make a dramatic difference in emissions.

- These changes can be healthy and fun: more exercise, more walking and cycling, more taking advantage of activities in our own backyards.
- You don't have to be doctrinaire; that is why I keep a budget rather than just saying no to any pleasure like the occasional Chinese dinner, or the drive to the cottage that eats up three days' worth of carbon—if I stay there all summer, I can amortize that over the season.
- Perhaps most importantly, even though I had been writing about it for years, I learned to look at everything through the lens of *sufficiency*.

I noted earlier that "Nobody buys a cup of oil or electricity; they buy what it does." Bart Hawkins Kreps, in an essay in his new book, *Energy Scarcity and Economic Sufficiency*, calls these energy services.

> [Elizabeth] Shove argues that it is important not to fixate on ways to provide a given energy service more efficiently because it may be equally if not more important to question whether it is a good idea to provide that service, in its widespread form, at all. She adds that a focus on energy efficiency is conducive to a "business as usual" orientation: "Programs of energy efficiency are politically uncontroversial precisely because they take current interpretations of 'service' for granted."[4]

A car is still a car, and a house is still a house, even if it is more efficient or uses a different form of energy; it doesn't change the way people actually live, or question whether they actually need the service at all. But we can't take anything for granted and just assume that hydrogen or small modular reactors or carbon capture and storage will save us. There is no fancy new technology that is going to dramatically cut our emissions in half by 2030 or take them to net zero by 2050, but we don't need drama. Instead, we have to change the way we live.

I noted early in the book that it is our consumption that matters, not production, and that the answer is to simply not buy what they are selling. But I have to acknowledge that this is not easy, when fossil fuels drive the economy, when there are no decent trains, when a parking permit costs a tenth the price of a transit pass, when we work long hours and have to race between everything, when so many of our politicians are in hock to the fossil fuel and real estate interests. They have all been so successful in making it extremely difficult to change.

It is also not easy in a society where there is so much pressure to consume, with so much status that comes from conspicuous consumption or even conspicuous conservation. The marketers and the politicians and the business world have been not only training us but also pleading with us to consume. After 9/11, George Bush told Americans to go shopping: "Get down to Disney World in Florida," he said. "Take your families and enjoy life, the way we want it to be enjoyed."[5] Go consume, that's what keeps the economy going.

Changing that is going to be hard, but it doesn't mean people have to be miserable, and it is much easier than it used to be. Jimmy Carter told us to turn down the thermostat and put on a sweater; very few people listened because people like to be warm and comfortable. Today, we can build our homes to standards like Passive House, the most comfortable buildings in the world; the walls are your sweater. We've got e-bikes that can whiz us around town. We have computers that can zip us to work. We have an electrical supply that is getting greener every day with cheap solar, wind, and batteries. We have induction stoves that cook better than gas. It's getting to the point that we can consciously decouple our daily lives from fossil fuels and live as happily as we ever did.

We don't have to actually change that much because living a lifestyle based on sufficiency isn't about doing without; it is about using less, about using better, and making that modal shift to alternatives.

While our individual actions matter, we cannot do it on our own; many people have to participate in these changes, they have to be societal as well as personal. We have seen that happen too.

A hundred and forty ago, spitting was common, people did it in the streets, in public buildings, really, just about everywhere. A few years after Robert Koch determined that it spread the tuberculosis bacillus, cities started banning spitting in public; New York City did in 1896. Nobody paid much attention until the Ladies' Health Protective Association (LHPA) got on the case. According to historian Felice Batlan, public action made a big difference, highlighting "importance of local, community action being done in the name of citizens who want to have and create a larger public good."[6]

Then there is the example we have discussed previously, smoking. This was a combination of scientific knowledge, political action, and societal change, all in the face of well-funded opposition much like we see now with climate change (and many of the same people). The smoking rate has dropped by 67% since 1965.

There is also research showing how people do care about what other people do, about social norms. Robert Cialdini did a famous study with those silly "Save the planet, reuse your towels" signs you see in hotel bathrooms. Some simply said "Help save the environment by reusing your towels during your stay." Others said "Almost 75% of guests who are asked to participate in our new resource savings program do help by using their towels more than once." With the second message, the rate of towel reuse went up over 25%.[7] It's not just because it's green, it's because your neighbors are doing it too.

With societal change, we eventually get political change, even if it is a trailing indicator. In the UK, Conservative Prime Minister Boris Johnson has promised radical decarbonization. In the US, President Joe Biden is rejoining the Paris Agreement. In Canada, Prime Minister Justin Trudeau is implementing a carbon tax with a rate high enough that it will make decarbonization an economic no-brainer for any household or business.

We may well look back and see that the pandemic of 2020 was the turning point. It disrupted so much, but it also shattered a lot of the conventional wisdom about the way we work, the way we communicate, even the way we are entertained. Much has changed, but not always for the worse; we can build on this.

Finally, we can never forget why our 1.5-degree lifestyle is based on 2.5 tonnes of CO_2 per year: it is the global carbon budget divided equally among everyone on Earth, divided fairly and equitably. It is sufficient. It gives everyone enough. It keeps the planet from cooking, and we can do it.

Notes

Introduction

1. "Global CO_2 Emissions in 2019," February 11, 2020, www.iea.org.

Chapter 1: What's the 1.5-Degree Lifestyle?

1. "End of Year Review," *One Tonne of Carbon per Year* (blog), 2020, rosalindreadhead.wordpress.com/journal.
2. Drijfhout, Sybren, et al., "Catalogue of Abrupt Shifts in Intergovernmental Panel on Climate Change Climate Models," *PNAS*, 112 (43), October 12, 2015, pnas.org.
3. Hausfather, Zeke, "Analysis: Why the IPCC 1.5C Report Expanded the Carbon Budget," October, 8, 2018, carbonbrief.org.
4. Lahn, Bård, "A History of the Global Carbon Budget," January 5, 2020, onlinelibrary.wiley.com.
5. Ibid.
6. Marvel, Kate, "Thinking about Climate on a Dark, Dismal Morning," *Scientific American*, December 25, 2018, blogs.scientificamerican.com.

Chapter 2: Equity, Fairness, and the 2.5-Tonne Budget

1. Roser, Max, "The World's Energy Problem," *Our World in Data*, December 10, 2020.
2. Peters, Glen P., and Hertwich, Edgar G., "CO_2 Embodied in International Trade with Implications for Global Climate Policy," *Environmental Science & Technology*, 42 (5), pp. 1401–7, 2008.
3. Akenji, Lewis, et al., *1.5-Degree Lifestyles: Targets and Options for Reducing Lifestyle Carbon Footprints*, Institute for Global Environmental Strategies, February 2019, iges.or.
4. Ibid.
5. Berners-Lee, Mike, *How Bad Are Bananas? The Carbon Footprint of Everything*, Profile Books, May 13, 2010, profilebooks.com.
6. Ibid.
7. Ritchie, Hannah, and Roser, Max, "Environmental Impacts of Food Production," January 2020, ourworldindata.org.

8. Poore, J., and Nemecek, T., "Reducing Food's Environmental Impacts Through Producers and Consumers," *Science*, V. 360 (6392), pp. 987–92, June 2018.

Chapter 3: Why Individual Actions Matter

1. Zhou, Li, "Elizabeth Warren Blasts the Plastic Straw Debate as a Fossil Fuel Industry Distraction Tactic," *Vox*, September 5, 2019, vox.com.
2. Griffin, Paul, July 2017, *The Carbon Majors Database: CDP Carbon Majors Report 2017*, cdp.net.
3. "2020 Democrats at Climate Town Hall," *New York Times*, updated June 9, 2020.
4. Riley, Tess, "Just 100 Companies Responsible for 71% of Global Emissions, Study Says," *Guardian*, July 10, 2017.
5. Murray, James, "Peabody Energy's Coal Mine Write-down 'Speaks Volumes About the State of the Industry,'" *NS Energy*, August 7, 2020, nsenergybusiness.com.
6. Grandoni, Dino, "Big Oil Just Isn't as Big as It Once Was," *Washington Post*, September 4, 2020.
7. Griffin, P., *The Carbon Majors Database*, cdp.net.
8. Packard, Vance, *The Waste Makers*, Longmans, London, 1960.
9. Grover, Sami, "In Defense of Eco-Hypocrisy, Again," *Treehugger*, updated March 22, 2019.
10. Mann, Michael E., "Lifestyle Changes Aren't Enough to Save the Planet. Here's What Could," *Time*, September 12, 2019.
11. Alter, Lloyd, "Recycling Is BS; Make Nov. 15 Zero Waste Day, Not America Recycles Day," *Treehugger*, updated October 11, 2018.
12. Jarvis, David, "We Blame Wrong Industries for Pollution," *Change Incorporated*, August 19, 2020, changeincorporated.com.
13. Mann, "Lifestyle Changes."
14. Hackel, Leor, and Sparkman, Gregg, "Reducing Your Carbon Footprint Still Matters," *Slate*, October 26, 2018.
15. Brandt, Allan, "Blow Some My Way: Passive Smoking, Risk and American Culture," in *Ashes to Ashes: The History of Smoking and Health*, Stephen Lock, Lois Reynolds, and E. M. Tansey, Eds., Amsterdam: Rodopi, 1998, pp. 164–191, columbia.edu.
16. Shrimsley, Robert, "Once You're Accused of Virtue-signalling, You Can't Do Anything Right," *Financial Times*, May 10, 2019.
17. Ibid.
18. Karakaya, E., Yılmaz, B., and Alataş, S., "How Production-based and Consumption-based Emissions Accounting Systems Change Climate

Policy Analysis: The Case of CO_2 Convergence," *Environmenal Science and Pollution Research*, 26, pp. 16682–694, 2019, link.springer.com.
19. World Green Building Council, "Bringing Embodied Carbon Upfront: Embodied Carbon Call to Action Report," n.d., worldgbc.org.
20. *Economist*, "Carbon-reduction Targets: The Right Way to Measure Carbon Emissions," October 19, 2019.
21. Ibid.
22. C40 Cities, *The Future of Urban Consumption in a 1.5°C World*, c40.org/consumption.
23. Ibid.
24. Ibid.
25. Steffen, Alex, "My Other Car Is a Bright Green City," *Medium*, May 7, 2020, medium.com/@AlexSteffen.
26. Ibid.

Chaper 4: Energy, Efficiency, and Sufficiency

1. Smil, Vaclav, *Energy and Civilization: A History*, MIT Press, 2017.
2. Ibid.
3. Smil, Vaclav, *Growth: From Microorganisms to Megacities*, MIT Press, 2019.
4. Ibid.
5. Kormann, Caroline, "The False Choice Between Economic Growth and Combatting Climate Change," *New Yorker*, February 4, 2019.
6. Smil, *Growth*, p. 491.
7. Hickel, Jason, *Less Is More: How Degrowth Will Save the World*, Penguin Random House, 2020, jasonhickel.org.
8. Thomas, Katherine Woodward, *Conscious Uncoupling: 5 Steps to Living Happily Even After*, Harmony, 2016, consciousuncoupling.com.
9. Alexander, Samuel, "Life in a 'Degrowth' Economy, and Why You Might Actually Enjoy It," *The Conversation*, October 1, 2014.
10. Ibid.
11. Ibid.
12. Ibid.
13. Ivanova, D., and Wood, R., "The Unequal Distribution of Household Carbon Footprints in Europe and Its Link to Sustainability." *Global Sustainability*, 3, 2020, cambridge.org.
14. Wiedmann, T., et al., "Scientists' Warning on Affluence," *Nature Communications*, V. 11, 2020, nature.com.
15. De Decker, Kris, "Bedazzled by Energy Efficiency," *Low-tech Magazine*, January 9, 2018.

16. Linnanen, Lassi, *The Sufficiency Perspective in Climate Policy: How to Recompose Consumption*, Finnish Climate Change Panel, September 30, 2020, researchportal.helsinki.fi/en/publications.
17. Ibid.
18. Alexander, Samuel, "Introduction to 'Sufficiency Economy,'" *Simplicity Collective*, August 11, 2015.
19. Rams, Dieter, "The Power of Good Design," *Vitsoe*.

Chapter 5: What We Eat

1. Ritchie, Hannah, "You Want to Reduce the Carbon Footprint of Your Food? Focus on What You Eat, Not Whether Your Food Is Local," *Our World in Data*, January 24, 2020.
2. Rodrigue, Jean-Paul, and Notteboom, Theo, *The Geography of Transport Systems*, New York: Routledge, 5th edition, 2020, transportgeography.org.
3. Ritchie, "You Want to Reduce the Carbon Footprint of Your Food?"
4. NRDC, "What's at Stake: Up to 40 Percent of the Food in the United States Is Never Eaten," n.d., nrdc.org.
5. Dobbs, Richard, et al., *Resource Revolution: Meeting the World's Energy, Materials, Food, and Water Needs*, McKinsey and Co., November 1, 2011, mckinsey.com.
6. FoodPrint, "The Problem of Food Waste," n.d., note 52, foodprint.org.
7. Lehner, Peter, "Tackling Food Waste at Home," NRDC, August 21, 2012.
8. British Heart Foundation, "Portion Distortion: How Much Are We Really Eating?" 2013, bhf.org.uk/-/media/files/publications.
9. Nestle, Marion, "What to Eat," North Point Press, 2006, foodpolitics.com.
10. McDermott, Robyn, "The Carbon Footprints of Obesity, Chronic Disease and Population Growth: Four Things Doctors Can Do," *Medical Journal of Australia*, 192 (9), pp. 531–2, May 2010.
11. Jha, Alok, "Carbon Emissions Fuelled by High Rates of Obesity," *Guardian*, April 20, 2009.
12. Obesity Society, "Study Suggests Obesity Associated with Greater Greenhouse Gas Emissions," news release, December 20, 2019, eurekalert.org.
13. Magkos, Faidon, et al., "The Environmental Foodprint of Obesity," *Obesity*, V. 28 (1), pp. 73–9, January 2020.
14. Wilson, Bee, et al., "Our Gigantic Problem with Portions: Why Are We All Eating Too Much?" *Guardian*, April 25, 2016.
15. Nastu, Paul, "Carbon Footprint of Tropicana Orange Juice: 1.7 Kg," *Environment + Energy Leader*, January 23, 2009, environmentalleader.com.

16. Barach, Paul, "The Tragedy of Fritz Haber: The Monster Who Fed the World," *Mission.org*, August 2, 2016, medium.com.
17. Royal Society, "Ammonia: Zero-carbon Fertiliser, Fuel and Energy Store," February 19, 2020, royalsociety.org.
18. Brasington, Louis, "Hydrogen in China," Cleantech Group, September 24, 2019, cleantech.com.
19. LeCompte, Celeste, "Fertilizer Plants Spring Up to Take Advantage of U.S.'s Cheap Natural Gas," *Scientific American*, April 25, 2013.
20. Smil, Vaclav, *Energy and Civilization: A History*, MIT Press, 2017.
21. Reed, Roy, "Organic Farms Found Efficient," *New York Times*, July 20, 1975.
22. Muller, A., et al. "Strategies for Feeding the World More Sustainably with Organic Agriculture. *Nature Communications*, 8 (1290), November 14, 2017, nature.com.
23. Lynas, Mark, "Organic Farming Can Feed the World—Until You Read the Small Print," *Alliance for Science*, November 22, 2017, allianceforscience .cornell.edu.
24. Ibid.
25. Clark, Michael A., et al., "Global Food System Emissions Could Preclude Achieving the 1.5° and 2°C Climate Change Targets," *Science*, V. 370 (6517), November 6, 2020, pp. 705–8.
26. Ibid.
27. Author correspondence with Michael Clark, November 11, 2020.
28. Clark, "Global Food System Emissions."
29. Poore, J., and Nemecek, T., "Reducing Food's Environmental Impacts Through Producers and Consumers," *Science*, V. 360 (6392), June 1, 2018, pp. 987–92.
30. Ritchie, "You Want to Reduce the Carbon Footprint of Your Food?" note 2, ourworldindata.org.
31. Weber, Christopher, and Matthews, H. Scott, "Food-Miles and the Relative Climate Impacts of Food Choices in the United States," *Environmental Science & Technology*, 42 (10), 2008, pp. 3508–13.
32. Ibid.
33. Gallo, Andrea, et al., "Designing Sustainable Cold Chains for Long-Range Food Distribution: Energy-Effective Corridors on the Silk Road Belt," *Sustainability*, 9 (11), p. 2044, 2017, mdpi.com.
34. Rodrigue and Notteboom, *The Geography of Transport Systems*.
35. Ibid.
36. Ibid.
37. Ibid.

38. Tassou, S. A., De-Lille, G., and Ge, Y. T., "Food Transport Refrigeration: Approaches to Reduce Energy Consumption and Environmental Impacts of Road Transport," *Applied Thermal Engineering*, V. 29 (8–9), June 2009, pp. 1467–77, sciencedirect.com.
39. UK Government, "Calculate the Carbon Dioxide Equivalent Quantity of an F Gas," December 31, 2014, gov.uk/guidance.

Chapter 6: How We Live

1. Smil, Vaclav, *Growth: From Microorganisms to Megacities*, MIT Press, 2019.
2. Davies, Margery, *Woman's Place Is at the Typewriter: Office Work and Office Workers, 1870–1930*, Temple University Press, 1982, tupress.temple.edu.
3. Ibid.
4. Ibid.
5. Otto, Shawn Lawrence, *Fool Me Twice: Fighting the Assault on Science in America*, Rodale Books, 2011, penguinrandomhouse.com.
6. Erickson, Jennifer, "Top 10 U.S. Government Investments in 20th Century American Competitiveness," *Center for American Progress*, January 6, 2012, americanprogress.org.
7. Tobin, Kathleen, "The Reduction of Urban Vulnerability: Revisiting 1950s American Suburbanization as Civil Defence," *Cold War History*, 2 (2), pp. 1–32, January 2002.
8. Huck, Nichole, "'Passive Home' Movement a Success in Germany, but Not in Saskatchewan Where It Started," *CBC News*, posted August 5, 2015.
9. Alter, Lloyd, "Cities Need Goldilocks Housing Density: Not Too High or Low, But Just Right," *Guardian*, April 16, 2014.
10. Natural Resources Canada, "Urban Archetypes Project," February 11, 2020, nrcan.gc.ca.
11. Henry, Mike, "Harold Orr's Superinsulated Retrofits," *The Sustainable Home*, August 9, 2013, thesustainablehome.net.
12. Statista, "Volume of detached single-family homes in Canada from 2015 to 2023," 2021, statista.com.
13. Committee on Climate Change, *Net Zero: Technical Report*, May 2019, theccc.org.uk.
14. Mo, Weiwei, et al., "Embodied Energy Comparison of Surface Water and Groundwater Supply Options," *Water Research*, V. 45 (17), pp. 5577–86, 2011.
15. Melton, P., "The Embodied Energy of Tap Water," September 8, 2015, *BuildingGreen*, buildinggreen.com.
16. EPA, *Energy Efficiency in Water and Wastewater Facilities A Guide to Developing and Implementing Greenhouse Gas Reduction Programs*, 2015, epa.gov.

17. Melton, "The Embodied Energy of Tap Water."
18. Rockerfeller, Abby A., "Civilization & Sludge: Notes on the History of the Management of Human Excreta," *Current World Leaders*, V. 39 (6), pp. 3–18, 2009.
19. Ibid.
20. "Composting Toilets," *City Farmer*, Canada's Office of Urban Agriculture, cityfarmer.org.
21. Zimring, Carl, *Aluminum Upcycled: Sustainable Design in Historical Perspective*, Johns Hopkins University Press, 2017.
22. "Polyvinyl chloride," *Wikipedia*.
23. Alter, Lloyd, "Why Ground Source Heat Pumps Should Not Be Called Geothermal, Chapter CLXXI," *Treehugger*, August 16, 2020.
24. "Basics," *Passipedia: The Passive House Resource*, passipedia.org, June 12, 2020.
25. Sexton, Steven E., and Sexton, Alison L., "Conspicuous Conservation: The Prius Effect and Willingness to Pay for Environmental Bona Fides," April 21, 2011, berkeley.edu.
26. Partridge, Emily, "Achieving Zero Carbon," *Architype*, July 7, 2020, architype.co.uk.
27. Susanka, Sarah, *The Not So Big House: Making Room for What Really Matters*, Random House, 2007, susanka.com.
28. Stone, Philip, and Luchetti, Robert, "Your Office Is Where You Are," *Harvard Business Review*, V. 63 (2), pp 102–17, March/April 1985.
29. Altig, David, et al., "Firms Expect Working from Home to Triple," *Federal Reserve Bank*, May 28, 2020, frbatlanta.org/blogs.
30. Lister, Kate, "What Is Your Work-From-Home Forecast for After Covid-19?" *Global Workplace Analytics*, globalworkplaceanalytics.com.
31. Ibid.
32. Jacobs, Jane, *Dark Age Ahead*, Vintage Canada, 2005, penguinrandom house.ca.
33. "10 Principles of New Urbanism," Michigan Land Use Institute, April 27, 2006, mlui.org.
34. Whittle, Natalie, "Welcome to the 15-Minute City, *Financial Times*, July 16, 2020.
35. O'Sullivan, Feargus, "Paris Mayor: It's Time for a '15-Minute City,'" *Bloomberg CityLab*, February 18, 2020, bloomberg.com.
36. Bacevice, Peter, et al., "Reimagining the Urban Office," *Harvard Business Review*, August 14, 2020.
37. Meier, Allison C., "In Epidemics, the Wealthy Have Always Fled," *Daily Jstor*, April 2, 2020, daily.jstor.org.

38. Nikiforuk, Andre, "Italy's Lesson: Do Much COVID-19 Care Away from Hospitals," *Tyee*, March 23, 2020.

Chapter 7: How We Move

1. Hoekstra, Auke, "Correcting Misinformation About Greenhouse Gas Emissions of Electric Vehicles: Auke Hoekstra's Response to Damien Ernst's Calculations," *Innovation Origins*, March 21, 2019.
2. Hickel, Jason, *Less Is More: How Degrowth Will Save the World*, Penguin Random House, 2020, jasonhickel.org.
3. Ibid.
4. Yan, Zhang, Sun, Yilei, and Goh, Brenda, "Exclusive: Tesla in Talks to Use CATL's Cobalt-free Batteries in China-made Cars—Sources," *Reuters*, February 17, 2020.
5. Cai, Hao, et al., "Well-to-Wheels Greenhouse Gas Emissions of Canadian Oil Sands Products: Implications for U.S. Petroleum Fuels," *Environmental Science & Technology*, 49 (13), pp. 8219–27, 2015.
6. Mumford, Lewis, *The Highway and the City* (Routledge, 1964), quoted in Shoup, Donald, *Parking and the City*, Routledge, 1964.
7. Ben-Joseph, Eran, *ReThinking a Lot: The Design and Culture of Parking*, MIT Press, 2015.
8. Kreps, Bart Hawkins, "Energy Sprawl in the Renewable-Energy Sector: Moving to Sufficiency in a Post-Growth Era," *American Journal of Economics and Sociology*, V. 79, July 2020.
9. McQueen, Michael, MacArthur, John, and Cherry, Christopher, "The E-Bike Potential: Estimating Regional E-Bike Impacts on Greenhouse Gas Emissions," *Transportation Research Part D: Transport and Environment*, V. 87, October 2020, sciencedirect.com.
10. Stuart, Shabazz, Twitter, August 2020, twitter.com/ShabazzStuart.
11. Fyhri, Aslak, and Sundfør, Hanne Beate, "Do People Who Buy E-bikes Cycle More?" *Transportation Research Part D: Transport and Environment*, V. 86, September 2020, sciencedirect.com.
12. Wilde, Parke, "Opinion: Flying Less Should Be a High-priority Climate Action," *Ensia*, December 17, 2019.
13. Lee, D. S., et al., "The Contribution of Global Aviation to Anthropogenic Climate Forcing for 2000 to 2018," *Atmospheric Environment*, V. 244, 2021.
14. Pidcock, Roz, and Yeo, Sophie, "Analysis: Aviation Could Consume a Quarter pf 1.5C Carbon Budget by 2050," *Carbon Brief*, August 8, 2016.
15. "Air Travel's Sudden Collapse Will Reshape a Trillion-dollar Industry," *Economist*, August 1, 2020.

16. Wilde, "Opinion: Flying Less."
17. Cadell, Cate, "Beijing Proudly Unveils Mega-Airport Due to Open in 2019," *Reuters*, October 16, 2017.
18. Morris, Hugh, "How Many Planes Are There in the World," *Telegraph*, August 16, 2017.
19. Nevins, Joseph, "Flying Less: Reducing Academia's Carbon Footprint," *Academic Flying* (blog), January 27, 2021, academicflyingblog.wordpress .com.
20. Wynes, Seth, and Donner, Simon D., *Addressing Greenhouse Gas Emissions from Business-Related Air Travel at Public Institutions: A Case Study of the University of British Columbia*, Pacific Institute for Climate Solutions, July 2018, pics.uvic.ca.
21. Mann, Michael, "Lifestyle Changes Aren't Enough to Save the Planet. Here's What Could," *Time*, September 12, 2019.
22. Kalmus, Peter, "A Climate Scientist Who Decided Not to Fly," *Grist*, February 21, 2016.

Chapter 8: Why We Buy

1. Packard, Vance, *The Waste Makers*, Longmans, London, 1960.
2. Ibid.
3. Ibid.
4. Cichon, Steve, "Everything from This 1991 Radio Shack Ad You Can Now Do with Your Phone," *HuffPost*, January 16, 2014.
5. Packard, *The Waste Makers*.
6. Ibid.
7. Ibid.
8. Ibid.
9. Ibid.
10. Ritchie, Hannah, and Roser, Max, "Plastic Pollution," *Our World in Data*, September 2018.
11. "Our Planet Is Drowning in Plastic Pollution," n.d., unenvironment.org.
12. Rude, Emelyn, "What Take-Out Food Can Teach You About American History," *Time*, April 14, 2016.
13. Crowley, Carolyn Hughes, "Meet Me at the Automat: Horn & Hardart Gave Big City Americans a Taste of Good Fast Food in Its Chrome-and-Glass Restaurants," *Smithsonian Magazine*, August 2001.
14. "Transcript of President Dwight D. Eisenhower's Farewell Address (1961)," *Our Documents*, www.ourdocuments.gov.
15. Ibid.

16. Rude, "What Take-Out Food Can Teach You."
17. Newburger, Emma, and Lucas, Amelia, "Plastic Waste Surges as Coronavirus Prompts Restaurants to Use More Disposable Packaging," *CNBC*, June 28, 2020.
18. Guszkowski, Joe, "How Long Will the Delivery Boom Last?" *Restaurant Business Online*, August 11, 2020.
19. Zimring, Carl, *Aluminum Upcycled: Sustainable Design in Historical Perspective*, Johns Hopkins University Press, 2017.
20. "Tap Water in Toronto," *Toronto*, n.d., www.toronto.ca.
21. Royte, Elizabeth, *Bottlemania: How Water Went on Sale and Why We Bought It*, Bloomsbury, 2009.
22. Ibid.
23. Ibid.
24. Brown, Jessica, "How Much Water Should You Drink a Day?" *BBC Future*, May 1, 2020, bbc.com/future/story.
25. Hickel, Jason, *Less Is More: How Degrowth Will Save the World*, Penguin Random House, 2020, jasonhickel.org.
26. Zheng, J., and Suh, S., "Strategies to Reduce the Global Carbon Footprint of Plastics," *Nature Climate Change*, 9, pp. 374–78, 2019, nature.com.
27. Filella, Montserrat, "Antimony and PET Bottles: Checking Facts," *Chemosphere*, V. 261, December 2020, sciencedirect.com.
28. Carbon Tracker, "Oil Industry Betting Future on Shaky Plastics as World Battles Waste," September 4, 2020, carbontracker.org.
29. Ibid.
30. Keel, A. Tison, "The Economics of PET Recycling: Virgin PET Overcapacity Affects Demand for Recycled Resin," *Recycling Today*, February 2, 2017.
31. Ellen MacArthur Foundation, n.d., ellenmacarthurfoundation.org.
32. O'Dea, S., "Smartphones in the U.S.: Statistics & Facts," *statista*, January 26, 2021, statista.com.
33. Smil, Vaclav, "Embodied Energy: Mobile Devices and Cars [Numbers Don't Lie]," *IEEE Spectrum*, V. 53 (5), May 2016, ieeexplore.ieee.org.
34. Ibid.
35. Apple, *Environmental Progress Report 2020*, apple.com.
36. C40 Cities, *The Future of Urban Consumption in a 1.5°C World*, June 2019, c40.org/consumption.
37. Chinasamy, Jasmine, "'A Monstrous Disposable Industry': Fast Facts About Fast Fashion," *Unearthed*, September 12, 2019, unearthed.greenpeace.org.
38. World Bank, "How Much Do Our Wardrobes Cost to the Environment?" September 23, 2019, worldbank.org.

39. Ibid.
40. Morgan, Louise R., and Birtwistle, Grete, "An Investigation of Young Fashion Consumers' Disposal Habits," *International Journal of Consumer Studies*, V. 33 (2), pp. 190–8, March 2009.
41. Roberts-Islam, Brooke, "The True Cost of Brands Not Paying for Orders During the COVID-19 Crisis," *Forbes*, March 30, 2020.
42. "The 7 R's for Fashion Lovers," *Fashion Takes Action*, January 27, 2018.
43. Heathcote, Edwin, "The Curse of the Airbnb Aesthetic," *Financial Times*, August 21, 2020.
44. Goldschmidt, Debra, "Eighth Child Death from Fallen IKEA Dresser Prompts Recall Reminder," *CNN*, November 21, 2017.
45. Veblen, Thorstein, *The Theory of the Leisure Class: An Economic Study of Institutions*, Chapter 3: Conspicuous Leisure, New York: Macmillan, 1899.
46. Keynes, John Maynard, "Economic Possibilities for Our Grandchildren," *Panarchy*, panarchy.org.
47. Russell, Bertrand, "In Praise of Idleness," *Harper's*, October 1932.
48. Popova, Maria, "In Praise of Idleness: Bertrand Russell on the Relationship Between Leisure and Social Justice," *Brain Pickings*, n.d.
49. Russell, "In Praise of Idleness."
50. Freidman, Benjamin, "Work and Consumption in an Era of Unbalanced Technological Advance," *Journal of Evolutionary Economics*, V. 27, pp. 221–37, 2017.
51. Hickel, *Less Is More*.
52. Ibid.
53. International Labour Organization Newsroom, "Americans Work Longest Hours Among Industrialized Countries, Japanese Second Longest," September 6, 1999, ilo.org.
54. U.S. Bureau of Labor Statistics, "Economic News Release," June 25, 2020, bls.gov.
55. Harper, Lauren, "Cut! How the Entertainment Industry Is Reducing Environmental Impacts," *State of the Planet*, March 29, 2018, blogs.ei.columbia.edu.
56. Fitzpatrick, Kyle Raymond, "Behind Every Film Production Is a Mess of Environmental Wreckage," *Vice*, October 15, 2019.
57. Glance, David, "Universities Are Still Standing: The MOOC Revolution That Never Happened," *The Conversation*, July 14, 2014.
58. Ibid.
59. Steven Leacock, *My Financial Career and Other Follies*, 1895.
60. Gordon, Steven, "In the Future Everything Will Be a Coffee Shop," *Speculist*, December 26, 2011, blog.speculist.com.

Chapter 9: Conclusion: In Pursuit of Sufficiency

1. Boudette, Neal E., "Looking to Buy a Used Car in the Pandemic? So Is Everyone Else," *New York Times*, September 9, 2020.
2. Fischler, Marcelle Sussman, "Florida Attracts More Northerners," *New York Times*, September 4, 2020.
3. Guszkowski, Joe, "How Long Will the Delivery Boom Last?" *Restaurant Business*, August 11, 2020, restaurantbusinessonline.com.
4. Kreps, Bart Hawkins, "Energy Scarcity and Economic Sufficiency," *American Journal of Economics and Sociology*, V. 79, July 2020.
5. Fox, Justin, "Bush's Economic Mistakes: Telling Us to Go Shopping," *Time*, January 19, 2009.
6. Machemer, Theresa, "When a Women-led Campaign Made It Illegal to Spit in Public in New York City," *Smithsonian Magazine*, February 10, 2020.
7. Cialdini, Robert, "Don't Throw in the Towel: Use Social Influence Research," *Observer*, psychologicalscience.org/observer.

Index

About the Author

LLOYD ALTER is a writer, public speaker, and former architect, developer, and inventor. He has published over 14,000 articles on *Treehugger* and *Mother Nature Network* and has contributed to the *Guardian*, *Corporate Knights*, and *Azure* magazines. He has become convinced that we just use too much of everything—too much space, too much land, too much food, too much fuel, too much money—and that the key to sustainability is to simply use less, what he calls radical sufficiency. He teaches sustainable design at Ryerson School of Interior Design, and when not writing, he can often be found in his running shoes, on his bike, or in his 1989 Hudson single scull in Toronto, Canada.

Additional Resources from New Society Publishers

A Brief History of the Earth's Climate
Everyone's Guide to the Science of Climate Change
STEVEN EARLE

6 × 9" / 208 Pages
US/Can $19.99
PB ISBN 9780865719590

We're All Climate Hypocrites Now
How Embracing Our Limitations Can Unlock the Power of a Movement
SAMI GROVER

6 × 9" / 192 Pages
US/Can $19.99
PB ISBN 9780865719606

Facing the Climate Emergency
How to Transform Yourself with Climate Truth
MARGARET KLEIN SALAMON

5.5 × 8.5" / 160 Pages
US/Can $14.99
PB ISBN 9780865719415

How to Talk to Your Kids About Climate Change
Turning Angst into Action
HARRIET SHUGARMAN

6 × 9" / 192 Pages
US/Can $17.99
PB ISBN 9780865719361

For a full list of titles from New Society Publishers, visit newsociety.com

ABOUT NEW SOCIETY PUBLISHERS

New Society Publishers is an activist, solutions-oriented publisher focused on publishing books for a world of change. Our books offer tips, tools, and insights from leading experts in sustainable building, homesteading, climate change, environment, conscientious commerce, renewable energy, and more—positive solutions for troubled times.

DON'T EAT THIS BOOK *(but you could)*

We're proud to hold to the highest environmental and social standards of any publisher in North America. When you buy New Society books, you are part of the solution!

- We print all our books in North America, never overseas
- All our books are printed on **100% post-consumer recycled paper**, processed chlorine-free, with low-VOC vegetable-based inks (since 2002)
- Our corporate structure is an innovative employee shareholder agreement, so we're one-third employee-owned (since 2015)
- We're carbon-neutral (since 2006)
- We're certified as a B Corporation (since 2016)
- We're Signatories to the UN's Sustainable Development Goals (SDG) Publishers Compact (2020–2030, the Decade of Action)

At New Society Publishers, we care deeply about *what* we publish—but also about *how* we do business.

To download our full catalog, please visit newsociety.com/pages/nsp-catalogue.

Sign up for New Society Publishers' newsletter for information on upcoming titles, special offers, and author events (https://signup.e2ma.net/signup/1425175/42152/).

ENVIRONMENTAL BENEFITS STATEMENT

New Society Publishers saved the following resources by printing the pages of this book on chlorine free paper made with 100% post-consumer waste.

TREES	WATER	ENERGY	SOLID WASTE	GREENHOUSE GASES
29 FULLY GROWN	2,300 GALLONS	12 MILLION BTUs	100 POUNDS	12,610 POUNDS

Environmental impact estimates were made using the Environmental Paper Network Paper Calculator 4.0. For more information visit www.papercalculator.org